BRITANNICA Mathematics in Context

Made to Measure

Britannica

ENCYCLOPÆDIA BRITANNICA EDUCATIONAL CORPORATION

Mathematics in Context is a comprehensive curriculum for the middle grades. It was developed in collaboration with the Wisconsin Center for Education Research, School of Education, University of Wisconsin–Madison and the Freudenthal Institute at the University of Utrecht, The Netherlands, with the support of National Science Foundation Grant No. 9054928.

National Science Foundation

Opinions expressed are those of the authors
and not necessarily those of the Foundation

ISBN 0-7826-1515-5
1 2 3 4 5 WK 02 01 00 99 98

The *Mathematics in Context* Development Team

Mathematics in Context is a comprehensive curriculum for the middle grades. The National Science Foundation funded the National Center for Research in Mathematical Sciences Education at the University of Wisconsin–Madison to develop and field-test the materials from 1991 through 1996. The Freudenthal Institute at the University of Utrecht in The Netherlands, as a subcontractor, collaborated with the University of Wisconsin–Madison on the development of the curriculum.

The initial version of *Made to Measure* was developed by Jan de Lange and Monica Wijers. It was adapted for use in American schools by Aaron N. Simon and Mary C. Shafer.

National Center for Research in Mathematical Sciences Education Staff

Thomas A. Romberg
Director

Gail Burrill
Coordinator
Field Test Materials

Mary Ann Fix
Editorial Coordinator

James A. Middleton
Pilot Test Coordinator

Joan Daniels Pedro
Assistant to the Director

Margaret R. Meyer
Coordinator
Pilot Test Materials

Sherian Foster
Editorial Coordinator

Margaret A. Pligge
First Edition Coordinator

Project Staff

Jonathan Brendefur
Laura J. Brinker
James Browne
Jack Burrill
Rose Byrd
Peter Christiansen
Barbara Clarke
Doug Clarke
Beth R. Cole
Fae Dremock
Jasmina Milinkovic
Kay Schultz
Mary C. Shafer
Julia A. Shew
Aaron N. Simon
Marvin Smith
Stephanie Z. Smith
Mary S. Spence
Kathleen A. Steele

Freudenthal Institute Staff

Jan de Lange
Director

Els Feijs
Coordinator

Martin van Reeuwijk
Coordinator

Project Staff

Mieke Abels
Nina Boswinkel
Frans van Galen
Koeno Gravemeijer
Marja van den Heuvel-Panhuizen
Jan Auke de Jong
Vincent Jonker
Ronald Keijzer
Martin Kindt
Jansie Niehaus
Nanda Querelle
Anton Roodhardt
Leen Streefland
Adri Treffers
Monica Wijers
Astrid de Wild

Acknowledgments

Several school districts used and evaluated one or more versions of the materials: Ames Community School District, Ames, Iowa; Parkway School District, Chesterfield, Missouri; Stoughton Area School District, Stoughton, Wisconsin; Madison Metropolitan School District, Madison, Wisconsin; Milwaukee Public Schools, Milwaukee, Wisconsin; and Dodgeville School District, Dodgeville, Wisconsin. Two sites were involved in staff development as well as formative evaluation of materials: Culver City, California, and Memphis, Tennessee. Two sites were developed through partnership with Encyclopædia Britannica Educational Corporation: Miami, Florida, and Puerto Rico. University Partnerships were developed with mathematics educators who worked with preservice teachers to familiarize them with the curriculum and to obtain their advice on the curriculum materials. The materials were also used at several other schools throughout the United States.

We at Encyclopædia Britannica Educational Corporation extend our thanks to all who had a part in making this program a success. Some of the participants instrumental in the program's development are as follows:

Allapattah Middle School
Miami, Florida
Nemtalla (Nikolai) Barakat

Ames Middle School
Ames, Iowa
Kathleen Coe
Judd Freeman
Gary W. Schnieder
Ronald H. Stromen
Lyn Terrill

Bellerive Elementary
Creve Coeur, Missouri
Judy Hetterscheidt
Donna Lohman
Gary Alan Nunn
Jakke Tchang

Brookline Public Schools
Brookline, Massachusetts
Rhonda K. Weinstein
Deborah Winkler

Cass Middle School
Milwaukee, Wisconsin
Tami Molenda
Kyle F. Witty

Central Middle School
Waukesha, Wisconsin
Nancy Reese

Craigmont Middle School
Memphis, Tennessee
Sharon G. Ritz
Mardest K. VanHooks

Crestwood Elementary
Madison, Wisconsin
Diane Hein
John Kalson

Culver City Middle School
Culver City, California
Marilyn Culbertson
Joel Evans
Joy Ellen Kitzmiller
Patricia R. O'Connor
Myrna Ann Perks, Ph.D.
David H. Sanchez
John Tobias
Kelley Wilcox

Cutler Ridge Middle School
Miami, Florida
Lorraine A. Valladares

Dodgeville Middle School
Dodgeville, Wisconsin
Jacqueline A. Kamps
Carol Wolf

Edwards Elementary
Ames, Iowa
Diana Schmidt

Fox Prairie Elementary
Stoughton, Wisconsin
Tony Hjelle

Grahamwood Elementary
Memphis, Tennessee
M. Lynn McGoff
Alberta Sullivan

Henry M. Flagler Elementary
Miami, Florida
Frances R. Harmon

Horning Middle School
Waukesha, Wisconsin
Connie J. Marose
Thomas F. Clark

Huegel Elementary
Madison, Wisconsin
Nancy Brill
Teri Hedges
Carol Murphy

Hutchison Middle School
Memphis, Tennessee
Maria M. Burke
Vicki Fisher
Nancy D. Robinson

Idlewild Elementary
Memphis, Tennessee
Linda Eller

Jefferson Elementary
Santa Ana, California
Lydia Romero-Cruz

Jefferson Middle School
Madison, Wisconsin
Jane A. Beebe
Catherine Buege
Linda Grimmer
John Grueneberg
Nancy Howard
Annette Porter
Stephen H. Sprague
Dan Takkunen
Michael J. Vena

Jesus Sanabria Cruz School
Yabucoa, Puerto Rico
Andreíta Santiago Serrano

John Muir Elementary School
Madison, Wisconsin
Julie D'Onofrio
Jane M. Allen-Jauch
Kent Wells

Kegonsa Elementary
Stoughton, Wisconsin
Mary Buchholz
Louisa Havlik
Joan Olsen
Dominic Weisse

Linwood Howe Elementary
Culver City, California
Sandra Checel
Ellen Thireos

Mitchell Elementary
Ames, Iowa
Henry Gray
Matt Ludwig

New School of Northern Virginia
Fairfax, Virginia
Denise Jones

Northwood Elementary
Ames, Iowa
Eleanor M. Thomas

Orchard Ridge Elementary
Madison, Wisconsin
Mary Paquette
Carrie Valentine

Parkway West Middle School
Chesterfield, Missouri
Elissa Aiken
Ann Brenner
Gail R. Smith

Ridgeway Elementary
Ridgeway, Wisconsin
Lois Powell
Florence M. Wasley

Roosevelt Elementary
Ames, Iowa
Linda A. Carver

Roosevelt Middle
Milwaukee, Wisconsin
Sandra Simmons

Ross Elementary
Creve Coeur, Missouri
Annette Isselhard
Sheldon B. Korklan
Victoria Linn
Kathy Stamer

St. Joseph's School
Dodgeville, Wisconsin
Rita Van Dyck
Sharon Wimer

St. Maarten Academy
St. Peters, St. Maarten, NA
Shareed Hussain

Sarah Scott Middle School
Milwaukee, Wisconsin
Kevin Haddon

Sawyer Elementary
Ames, Iowa
Karen Bush Hoiberg

Sennett Middle School
Madison, Wisconsin
Brenda Abitz
Lois Bell
Shawn M. Jacobs

Sholes Middle School
Milwaukee, Wisconsin
Chris Gardner
Ken Haddon

Stephens Elementary
Madison, Wisconsin
Katherine Hogan
Shirley M. Steinbach
Kathleen H. Vegter

Stoughton Middle School
Stoughton, Wisconsin
Sally Bertelson
Polly Goepfert
Jacqueline M. Harris
Penny Vodak

Toki Middle School
Madison, Wisconsin
Gail J. Anderson
Vicky Grice
Mary M. Ihlenfeldt
Steve Jernegan
Jim Leidel
Theresa Loehr
Maryann Stephenson
Barbara Takkunen
Carol Welsch

Trowbridge Elementary
Milwaukee, Wisconsin
Jacqueline A. Nowak

W. R. Thomas Middle School
Miami, Florida
Michael Paloger

Wooddale Elementary Middle School
Memphis, Tennessee
Velma Quinn Hodges
Jacqueline Marie Hunt

Yahara Elementary
Stoughton, Wisconsin
Mary Bennett
Kevin Wright

Site Coordinators

Mary L. Delagardelle—Ames Community Schools, Ames, Iowa

Dr. Hector Hirigoyen—Miami, Florida

Audrey Jackson—Parkway School District, Chesterfield, Missouri

Jorge M. López—Puerto Rico

Susan Militello—Memphis, Tennessee

Carol Pudlin—Culver City, California

Reviewers and Consultants

Michael N. Bleicher
Professor of Mathematics
University of Wisconsin–Madison
Madison, WI

Diane J. Briars
Mathematics Specialist
Pittsburgh Public Schools
Pittsburgh, PA

Donald Chambers
Director of Dissemination
University of Wisconsin–Madison
Madison, WI

Don W. Collins
Assistant Professor of Mathematics Education
Western Kentucky University
Bowling Green, KY

Joan Elder
Mathematics Consultant
Los Angeles Unified School District
Los Angeles, CA

Elizabeth Fennema
Professor of Curriculum and Instruction
University of Wisconsin-Madison
Madison, WI

Nancy N. Gates
University of Memphis
Memphis, TN

Jane Donnelly Gawronski
Superintendent
Escondido Union High School
Escondido, CA

M. Elizabeth Graue
Assistant Professor of Curriculum and Instruction
University of Wisconsin–Madison
Madison, WI

Jodean E. Grunow
Consultant
Wisconsin Department of Public Instruction
Madison, WI

John G. Harvey
Professor of Mathematics and Curriculum & Instruction
University of Wisconsin–Madison
Madison, WI

Simon Hellerstein
Professor of Mathematics
University of Wisconsin–Madison
Madison, WI

Elaine J. Hutchinson
Senior Lecturer
University of Wisconsin–Stevens Point
Stevens Point, WI

Richard A. Johnson
Professor of Statistics
University of Wisconsin–Madison
Madison, WI

James J. Kaput
Professor of Mathematics
University of Massachusetts–Dartmouth
Dartmouth, MA

Richard Lehrer
Professor of Educational Psychology
University of Wisconsin–Madison
Madison, WI

Richard Lesh
Professor of Mathematics
University of Massachusetts–Dartmouth
Dartmouth, MA

Mary M. Lindquist
Callaway Professor of Mathematics Education
Columbus College
Columbus, GA

Baudilio (Bob) Mora
Coordinator of Mathematics & Instructional Technology
Carrollton-Farmers Branch Independent School District
Carrollton, TX

Paul Trafton
Professor of Mathematics
University of Northern Iowa
Cedar Falls, IA

Norman L. Webb
Research Scientist
University of Wisconsin–Madison
Madison, WI

Paul H. Williams
Professor of Plant Pathology
University of Wisconsin–Madison
Madison, WI

Linda Dager Wilson
Assistant Professor
University of Delaware
Newark, DE

Robert L. Wilson
Professor of Mathematics
University of Wisconsin–Madison
Madison, WI

BRITANNICA Mathematics in Context

Dear Teacher,

Welcome! *Mathematics in Context* is designed to reflect the National Council of Teachers of Mathematics Standards for School Mathematics and to ground mathematical content in a variety of real-world contexts. Rather than relying on you to explain and demonstrate generalized definitions, rules, or algorithms, students investigate questions directly related to a particular context and construct mathematical understanding and meaning from that context.

The curriculum encompasses 10 units per grade level. *Made to Measure* is designed to be the second unit in the geometry strand for grade 6/7, but it also lends itself to independent use—to help students learn about various measurement units for lengths, areas, volumes, and angles and use these units to measure common objects in real-life situations.

In addition to the Teacher Guide and Student Books, *Mathematics in Context* offers the following components that will inform and support your teaching:

- *Teacher Resource and Implementation Guide,* which provides an overview of the complete system, including program implementation, philosophy, and rationale
- *Number Tools,* Volumes 1 and 2, which are a series of blackline masters that serve as review sheets or practice pages involving number issues and basic skills
- *News in Numbers,* which is a set of additional activities that can be inserted between or within other units; it includes a number of measurement problems that require estimation.

Thank you for choosing *Mathematics in Context.* We wish you success and inspiration!

Sincerely,

The Mathematics in Context Development Team

Number Strand | Algebra Strand | Geometry Strand | Statistics Strand

Overview

BRITANNICA
Mathematics in Context
500 Kilometers
500 Miles
50°
40°
Reno
San Francisco
Denver
St. Louis
New York
Philadelphia
New Orleans
100°
90°
80°
26"
STEREO
The Netherlands

How to Use This Book

This unit is one of 40 for the middle grades. Each unit can be used independently; however, the 40 units are designed to make up a complete, connected curriculum (10 units per grade level). There is a Student Book and a Teacher Guide for each unit.

Each Teacher Guide comprises elements that assist the teacher in the presentation of concepts and in understanding the general direction of the unit and the program as a whole. Becoming familiar with this structure will make using the units easier.

Each Teacher Guide consists of six basic parts:

- Overview
- Student Materials and Teaching Notes
- Assessment Activities and Solutions
- Glossary
- Blackline Masters
- Try This! Solutions

Overview

Before beginning this unit, read the Overview in order to understand the purpose of the unit and to develop strategies for facilitating instruction. The Overview provides helpful information about the unit's focus, pacing, goals, and assessment, as well as explanations about how the unit fits with the rest of the *Mathematics in Context* curriculum.

Student Materials and Teaching Notes

This Teacher Guide contains all of the student pages, (except the Try This! activities), each of which faces a page of solutions, samples of students' work, and hints and comments about how to facilitate instruction. Note: Solutions for the Try This! activities can be found at the back of the Teacher Guide.

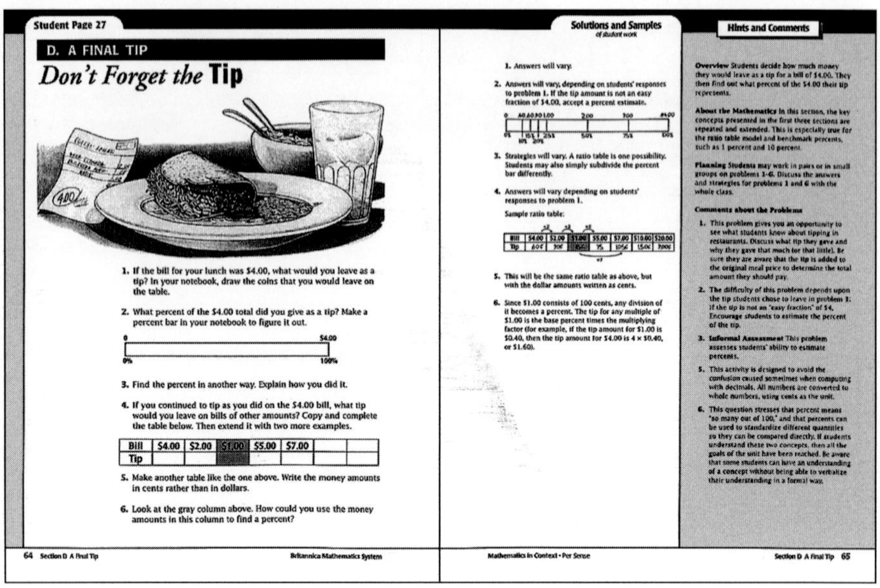

Student Page 27

D. A FINAL TIP

Don't Forget the **Tip**

1. If the bill for your lunch was $4.00, what would you leave as a tip? In your notebook, draw the coins that you would leave on the table.
2. What percent of the $4.00 total did you give as a tip? Make a percent bar in your notebook to figure it out.

0 — $4.00
0% — 100%

3. Find the percent in another way. Explain how you did it.
4. If you continued to tip as you did on the $4.00 bill, what tip would you leave on bills of other amounts? Copy and complete the table below. Then extend it with two more examples.

Bill	$4.00	$2.00	$1.00	$5.00	$7.00		
Tip							

5. Make another table like the one above. Write the money amounts in cents rather than in dollars.
6. Look at the gray column above. How could you use the money amounts in this column to find a percent?

64 Section D A Final Tip — Britannica Mathematics System

Solutions and Samples
of student work

1. Answers will vary.
2. Answers will vary, depending on students' responses to problem 1. If the tip amount is not an easy fraction of $4.00, accept a percent estimate.
3. Strategies will vary. A ratio table is one possibility. Students may also simply subdivide the percent bar differently.
4. Answers will vary depending on students' responses to problem 1.

Sample ratio table:

Bill	$4.00	$2.00	$1.00	$5.00	$7.00	$10.00	$20.00
Tip	60¢	30¢	[illegible]	75	105¢	150¢	300¢

5. This will be the same ratio table as above, but with the dollar amounts written as cents.
6. Since $1.00 consists of 100 cents, any division of it becomes a percent. The tip for any multiple of $1.00 is the base percent times the multiplying factor (for example, if the tip amount for $1.00 is $0.40, then the tip amount for $4.00 is 4 × $0.40, or $1.60).

Mathematics in Context • Per Sense

Hints and Comments

Overview Students decide how much money they would leave as a tip for a bill of $4.00. They then find out what percent of the $4.00 their tip represents.

About the Mathematics In this section, the key concepts presented in the first three sections are repeated and extended. This is especially true for the ratio table model and benchmark percents, such as 1 percent and 10 percent.

Planning Students may work in pairs or in small groups on problems 1–6. Discuss the answers and strategies for problems 1 and 6 with the whole class.

Comments about the Problems

1. This problem gives you an opportunity to see what students know about tipping in restaurants. Discuss what tip they gave and why they gave that much (or that little). Be sure they are aware that the tip is added to the original meal price to determine the total amount they should pay.
2. The difficulty of this problem depends upon the tip students chose to leave in problem 1. If the tip is not an "easy fraction" of $4, Encourage students to estimate the percent of the tip.
3. **Informal Assessment** This problem assesses students' ability to estimate percents.
5. This activity is designed to avoid the confusion caused sometimes when computing with decimals. All numbers are converted to whole numbers, using cents as the unit.
6. This question stresses that percent means "so many out of 100," and that percents can be used to standardize different quantities so they can be compared directly. If students understand these two concepts, then all the goals of the unit have been reached. Be aware that some students can have an understanding of a concept without being able to verbalize their understanding in a formal way.

Section D A Final Tip 65

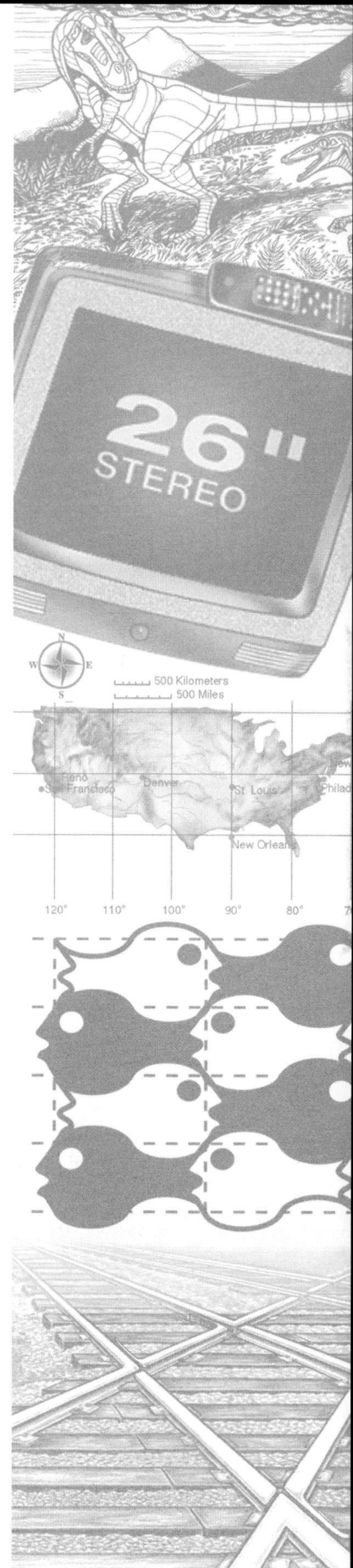

Each section within the unit begins with a two-page spread that describes the work students do, the goals of the section, new vocabulary, and materials needed, as well as providing information about the mathematics in the section and ideas for pacing, planning instruction, homework, and assessment.

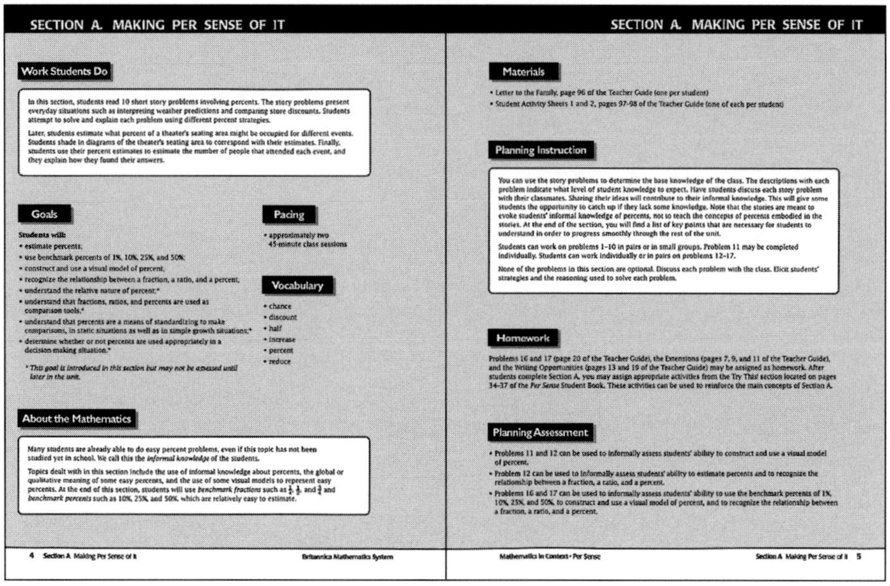

SECTION A. MAKING PER SENSE OF IT

Work Students Do

In this section, students read 10 short story problems involving percents. The story problems present everyday situations such as interpreting weather predictions and comparing store discounts. Students attempt to solve and explain each problem using different percent strategies.

Later, students estimate what percent of a theater's seating area might be occupied for different events. Students shade in diagrams of the theater's seating area to correspond with their estimates. Finally, students use their percent estimates to estimate the number of people that attended each event, and they explain how they found their answers.

Goals

Students will:
- estimate percents;
- use benchmark percents of 1%, 10%, 25%, and 50%;
- construct and use a visual model of percent;
- recognize the relationship between a fraction, a ratio, and a percent;
- understand the relative nature of percent;*
- understand that fractions, ratios, and percents are used as comparison tools;*
- understand that percents are a means of standardizing to make comparisons, in static situations as well as in simple growth situations;*
- determine whether or not percents are used appropriately in a decision-making situation.*

** This goal is introduced in this section but may not be assessed until later in the unit.*

Pacing

- approximately two 45-minute class sessions

Vocabulary

- chance
- discount
- half
- increase
- percent
- reduce

About the Mathematics

Many students are already able to do easy percent problems, even if this topic has not been studied yet in school. We call this the *informal knowledge* of the students.

Topics dealt with in this section include the use of informal knowledge about percents, the global or qualitative meaning of some easy percents, and the use of some visual models to represent easy percents. At the end of this section, students will use *benchmark fractions* such as $\frac{1}{4}$, $\frac{1}{2}$, and $\frac{3}{4}$ and *benchmark percents* such as 10%, 25%, and 50%, which are relatively easy to estimate.

4 Section A Making Per Sense of It — Britannica Mathematics System

SECTION A. MAKING PER SENSE OF IT

Materials

- Letter to the Family, page 96 of the Teacher Guide (one per student)
- Student Activity Sheets 1 and 2, pages 97–98 of the Teacher Guide (one of each per student)

Planning Instruction

You can use the story problems to determine the base knowledge of the class. The descriptions with each problem indicate what level of student knowledge to expect. Have students discuss each story problem with their classmates. Sharing their ideas will contribute to their informal knowledge. This will give some students the opportunity to catch up if they lack some knowledge. Note that the stories are meant to evoke students' informal knowledge of percents, not to teach the concepts of percents embodied in the stories. At the end of the section, you will find a list of key points that are necessary for students to understand in order to progress smoothly through the rest of the unit.

Students can work on problems 1–10 in pairs or in small groups. Problem 11 may be completed individually. Students can work individually or in pairs on problems 12–17.

None of the problems in this section are optional. Discuss each problem with the class. Elicit students' strategies and the reasoning used to solve each problem.

Homework

Problems 16 and 17 (page 20 of the Teacher Guide), the Extensions (pages 7, 9, and 11 of the Teacher Guide), and the Writing Opportunities (pages 13 and 19 of the Teacher Guide) may be assigned as homework. After students complete Section A, you may assign appropriate activities from the Try This! section located on pages 34–37 of the *Per Sense* Student Book. These activities can be used to reinforce the main concepts of Section A.

Planning Assessment

- Problems 11 and 12 can be used to informally assess students' ability to construct and use a visual model of percent.
- Problem 12 can be used to informally assess students' ability to estimate percents and to recognize the relationship between a fraction, a ratio, and a percent.
- Problems 16 and 17 can be used to informally assess students' ability to use the benchmark percents of 1%, 10%, 25%, and 50%, to construct and use a visual model of percent, and to recognize the relationship between a fraction, a ratio, and a percent.

Mathematics in Context • Per Sense — Section A Making Per Sense of It 5

Assessment Activities and Solutions

Information about assessment can be found in several places in this Teacher Guide. General information about assessment is given in the Overview; informal assessment opportunities are identified on the teacher pages that face each student page; and the Assessment Activities section of this guide provides formal assessment opportunities.

Glossary

The Glossary defines all vocabulary words listed on the Section Opener pages. It includes mathematical terms that may be new to students, as well as words associated with the contexts introduced in the unit. (Note: The Student Book does not have a glossary. This allows students to construct their own definitions, based on their personal experiences with the unit activities.)

Blackline Masters

At the back of this Teacher Guide are blackline masters for photocopying. The blackline masters include a letter to families (to be sent home with students before beginning the unit), several student activity sheets, and assessment masters.

Try This! Solutions

Also included in the back of this Teacher Guide are the solutions to several Try This! activities—one related to each section of the unit—that can be used to reinforce the unit's main concepts. The Try This! activities are located in the back of the Student Book.

Unit Focus

This unit covers the measurement of length, area, volume, and angle. Throughout the unit, students take and estimate measurements using parts of the human body as measuring units. The unit begins with historic units of measure based upon parts of the human body and gradually moves to current measurement units of length, area, volume, and angle. Students learn to appreciate the importance of standard measurement units and the connection between different units within and across measurement systems.

Mathematical Content

- estimating and calculating lengths, areas, volumes, and angles
- using the relationships between and within the metric and English measurement systems
- recognizing, selecting, and using metric and English units of measurement for length, area, volume, and angle
- using formulas and rules of thumb to estimate and calculate lengths, areas, volumes, and angles
- understanding and using the relationships between length, area, volume, and angle

Prior Knowledge

This unit assumes that students have an understanding of the following:

- the difference between length, area, and volume
- how to use rulers to measure lengths in metric and English units
- how to find the areas of rectangular shapes in square units
- units of measure used for liquid volume (liter, milliliter, pint, and so on)
- how to measure angles using a compass card or a protractor

It would be helpful if students are already familiar with units of measure used for solid volume, such as cubic centimeters.

Planning and Preparation

Pacing: 16 days

Section	Work Students Do	Pacing*	Materials
A. Lengths	■ measure lengths of objects using pace, fathom, and thumb ■ find personal values for measures ■ investigate standard measurement units ■ discover a relationship between foot length and shoe size	4 days	■ Letter to the Family (one per student) ■ centimeter and inch rulers (one per student) ■ 5-meter piece of rope (one per class) ■ inch or centimeter tape measures (one per student) ■ world map, optional (one per class) ■ pennies (10 per student)
B. Areas	■ measure their body surface areas using various methods	4 days	■ See page 31 of the Teacher Guide for a complete list of the materials and quantities needed.
C. Volumes	■ measure volumes by working with the volumes of solid objects ■ study the metric units of volume by examining boxes ■ estimate the volumes of various liquid containers in liquid volume units	4 days	■ centimeter grid paper (one sheet per student) ■ cubic decimeter made from a net, optional (one per class) ■ tissue boxes (one per group) ■ centimeter and inch rulers (one per student) ■ See page 49 of the Teacher Guide for a complete list of the materials and quantities needed.
D. Angles	■ measure angles through studying ergonomics ■ investigate the angles in hand movements and sight and in equipment related to the hand (a keyboard) and sight (a computer monitor) ■ investigate angles involved in the back and arm and in furniture related to them (chairs)	4 days	■ dictionary or encyclopedia (one per class) ■ $8\frac{1}{2}''\times 11''$ paper (one sheet per student) ■ compass cards or protractors (one per student) ■ string, optional (one piece about 1.5 m long per group) ■ pens, optional (one per group) ■ drawing chalk or sheets of paper about 2 square meters, optional (one per group)

* One day is approximately equivalent to one 45-minute class session.

Preparation

In the *Teacher Resource and Implementation Guide* is an extensive description of the philosophy underlying both the content and the pedagogy of the *Mathematics in Context* curriculum. Suggestions for preparation are also given in the Hints and Comments columns of this Teacher Guide. You may want to consider the following:

- Work through the unit before teaching it. If possible, take on the role of the student and discuss your strategies with other teachers.
- Use the overhead projector for student demonstrations, particularly with overhead transparencies of the student activity sheets and any manipulatives used in the unit.
- Invite students to use drawings and examples to illustrate and clarify their answers.
- Allow students to work at different levels of sophistication. Some students may need concrete materials, while others can work at a more abstract level.
- Provide opportunities and support for students to share their strategies, which often differ. This allows students to take part in class discussions and introduces them to alternative ways to think about the mathematics in the unit.
- In some cases, it may be necessary to read the problems to students or to pair students to facilitate their understanding of the printed materials.
- A list of the materials needed for this unit is in the chart on page xiii.
- Try to follow the recommended pacing chart on page xiii. You can easily spend more time on this unit than the number of class periods indicated. Bear in mind, however, that many of the topics introduced in this unit will be revisited and covered more thoroughly in other *Mathematics in Context* units.

Resources

For Teachers	For Students
Books and Magazines • *Mathematics Assessment: Myths, Models, Good Questions, and Practical Suggestions,* edited by Jean Kerr Stenmark (Reston, Virginia: The National Council of Teachers of Mathematics, Inc., 1991) • Other good resources are Etienne Grandjean's *Fitting the Task to the Man,* 4th ed. (Washington, D.C.: Taylor & Francis, Inc., 1988) and Isaac Asimov's *Realm of Measure* (Boston, MA: Houghton Mifflin Company, 1960).	**Videos** Mathsphere Videos • *High Hopes* • *If the Shoe Fits* • *A Drop in the Ocean* • *For Good Measure* • *Level Pegging* • *An Area of Concern* (available from Encyclopædia Britannica)

Assessment

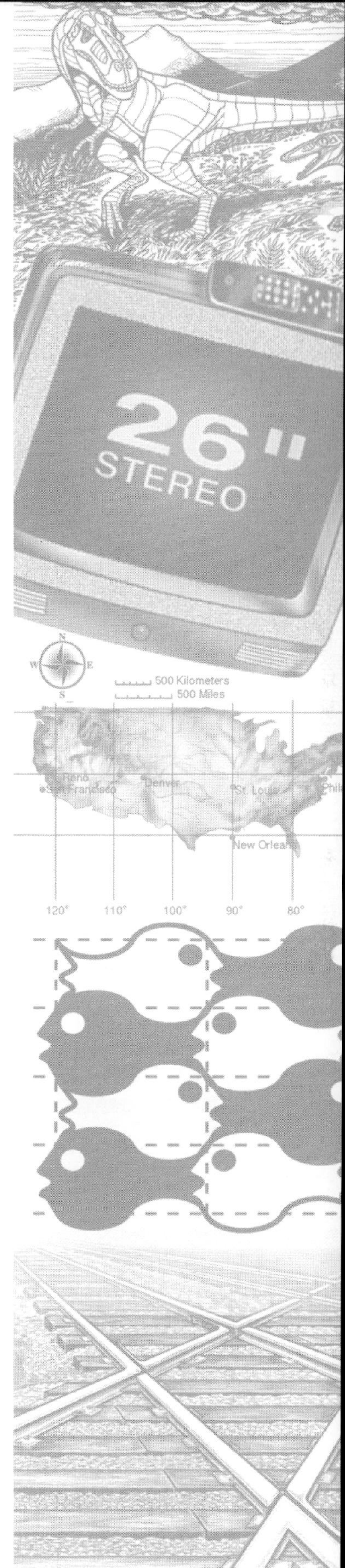

Planning Assessment

In keeping with the NCTM Assessment Standards, valid assessment should be based on evidence drawn from several sources. (See the full discussion of assessment philosophies in the *Teacher Resource and Implementation Guide.*) An assessment plan for this unit may draw from the following sources:

- Observations—look, listen, and record observable behavior.
- Interactive Responses—in a teacher-facilitated situation, note how students respond, clarify, revise, and extend their thinking.
- Products—look for the quality of thought evident in student projects, test answers, worksheet solutions, or writings.

These categories are not meant to be mutually exclusive. In fact, observation is a key part of assessing interactive responses and also key to understanding the end results of projects and writings.

Ongoing Assessment Opportunities

- **Problems within Sections**
  To evaluate ongoing progress, *Mathematics in Context* identifies informal assessment opportunities and the goals that these particular problems assess throughout the Teacher Guide. There are also indications as to what you might expect from your students.

- **Section Summary Questions**
  The Summary questions at the end of each section are vehicles for informal assessment (see Teacher Guide pages 28, 46, 64, and 78).

End-of-Unit Assessment Opportunities

In the back of this Teacher Guide, there is a set of assessment problems that can be given at the end of the unit. It will take approximately one 45-minute class session. There is also an alternative assessment task, which may be done by students over a longer period of time and can be completed at home. For a more detailed description of the assessment activity, see the Assessment Overview (Teacher Guide pages 80 and 81).

You may also wish to design your own culminating project or let students create one that will tell you what they consider is important in the unit. For more assessment ideas, refer to the charts on pages xvi and xvii.

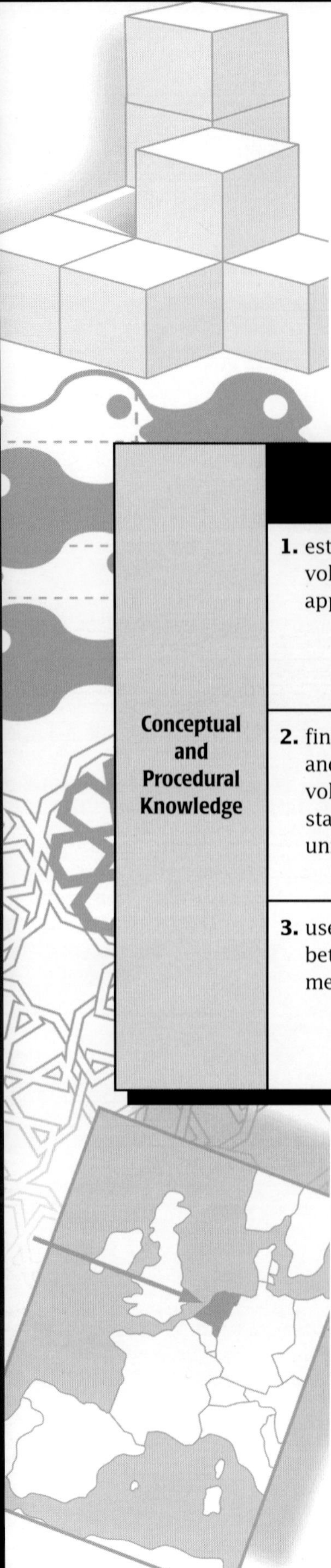

Goals and Assessment

In the *Mathematics in Context* curriculum, unit goals, categorized according to cognitive procedures, relate to the strand goals and the NCTM Curriculum and Evaluation Standards. Additional information about these goals is found in the *Teacher Resource and Implementation Guide.* The *Mathematics in Context* curriculum is designed to help students develop their abilities so that they can perform with understanding in each of the categories listed below. It is important to note that the attainment of goals in one category is not a prerequisite to attaining those in another category. In fact, students should progress simultaneously toward several goals in different categories.

	Goal	Ongoing Assessment Opportunities	End-of-Unit Assessment Opportunities
Conceptual and Procedural Knowledge	**1.** estimate lengths, areas, volumes, and angles using appropriate units	**Section A** p. 28, #31 **Section C** p. 64, #20 **Section D** p. 72, #3 p. 78, #10	Made to Measure Problems, p. 94, #1a, #1e p. 95, #3b p. 96, #5c Made to Measure Project, p. 97
	2. find (compute, measure, and draw) lengths, areas, volumes, and angles using standard and nonstandard units	**Section A** p. 28, #31 **Section C** p. 64, #20 **Section D** p. 72, #3 p. 76, #9 p. 78, #10	Made to Measure Problems, p. 94, #1a p. 95, #2, #3a p. 96, #4a, #5a, #5b Made to Measure Project, p. 97
	3. use simple relationships between and within measurement systems	**Section C** p. 64, #21	Made to Measure Problems, p. 94, #1d p. 95, #2 p. 96, #5b Made to Measure Project, p. 97

	Goal	Ongoing Assessment Opportunities	End-of-Unit Assessment Opportunities
Reasoning, Communicating, Thinking, and Making Connections	**4.** recognize that different units of measurement and different measurement systems exist	**Section A** p. 28, #31 **Section B** p. 46, #16 **Section C** p. 64, #20	Made to Measure Problems, p. 94, #1c p. 95, #2 Made to Measure Project, p. 97
	5. recognize the importance of standard (measurement) units	**Section B** p. 46, #16	Made to Measure Problems, p. 94, #1c p. 95, #2 Made to Measure Project, p. 97
	6. understand and use the relationships between length, area, volume, and angle	**Section C** p. 64, #20	Made to Measure Problems, p. 94, #1e p. 95, #3b, #3c p. 96, #4b Made to Measure Project, p. 97

	Goal	Ongoing Assessment Opportunities	End-of-Unit Assessment Opportunities
Modeling, Nonroutine Problem-Solving, Critically Analyzing, and Generalizing	**7.** use geometric models to solve problems	**Section D** p. 78, #10	Made to Measure Problems, p. 94, #3a, #3b, #3c Made to Measure Project, p. 97
	8. use formulas and rules of thumb to estimate and calculate length, area, and volume	**Section B** p. 46, #16 **Section C** p. 64, #20	Made to Measure Problems, p. 94, #1e p. 95, #3b, #3c Made to Measure Project, p. 97
	9. develop and use a frame of reference for measurement to solve problems	**Section B** p. 46, #16	Made to Measure Problems, p. 94, #1f p. 95, #2 p. 96, #5c Made to Measure Project, p. 97
	10. select and use appropriate measurement units	**Section A** p. 28, #31 **Section B** p. 46, #16 **Section C** p. 64, #20	Made to Measure Problems, p. 94, #1b, #1e p. 96, #5b Made to Measure Project, p. 97

More about Assessment

Scoring and Analyzing Assessment Responses

Students may respond to assessment questions with various levels of mathematical sophistication and elaboration. Each student's response should be considered for the mathematics that it shows, and not judged on whether or not it includes an expected response. Responses to some of the assessment questions may be viewed as either correct or incorrect, but many answers will need flexible judgment by the teacher. Descriptive judgments related to specific goals and partial credit often provide more helpful feedback than percentage scores.

Openly communicate your expectations to all students, and report achievement and progress for each student relative to those expectations. When scoring students' responses try to think about how they are progressing toward the goals of the unit and the strand.

Student Portfolios

Generally, a portfolio is a collection of student-selected pieces that is representative of a student's work. A portfolio may include evaluative comments by you or by the student. See the *Teacher Resource and Implementation Guide* for more ideas on portfolio focus and use.

A comprehensive discussion about the contents, management, and evaluation of portfolios can be found in *Mathematics Assessment: Myths, Models, Good Questions, and Practical Suggestions,* pp. 35–48.

Student Self-Evaluation

Self-evaluation encourages students to reflect on their progress in learning mathematical concepts, their developing abilities to use mathematics, and their dispositions toward mathematics. The following examples illustrate ways to incorporate student self-evaluations as one component of your assessment plan.

- Ask students to comment, in writing, on each piece they have chosen for their portfolios and on the progress they see in the pieces overall.
- Give a writing assignment entitled "What I Know Now about [a math concept] and What I Think about It." This will give you information about each student's disposition toward mathematics as well as his or her knowledge.
- Interview individuals or small groups to elicit what they have learned, what they think is important, and why.

Suggestions for self-inventories can be found in *Mathematics Assessment: Myths, Models, Good Questions, and Practical Suggestions,* pp. 55–58.

Summary Discussion

Discuss specific lessons and activities in the unit—what the student learned from them and what the activities have in common. This can be done in whole-class discussions, small groups, or in personal interviews.

Connections across the *Mathematics in Context* Curriculum

Made to Measure is the fourth unit in the geometry strand. The map below shows the complete *Mathematics in Context* curriculum for grade 6/7. This indicates where the unit fits in the geometry strand and in the overall picture.

A detailed description of the units, the strands, and the connections in the *Mathematics in Context* curriculum can be found in the *Teacher Resource and Implementation Guide.*

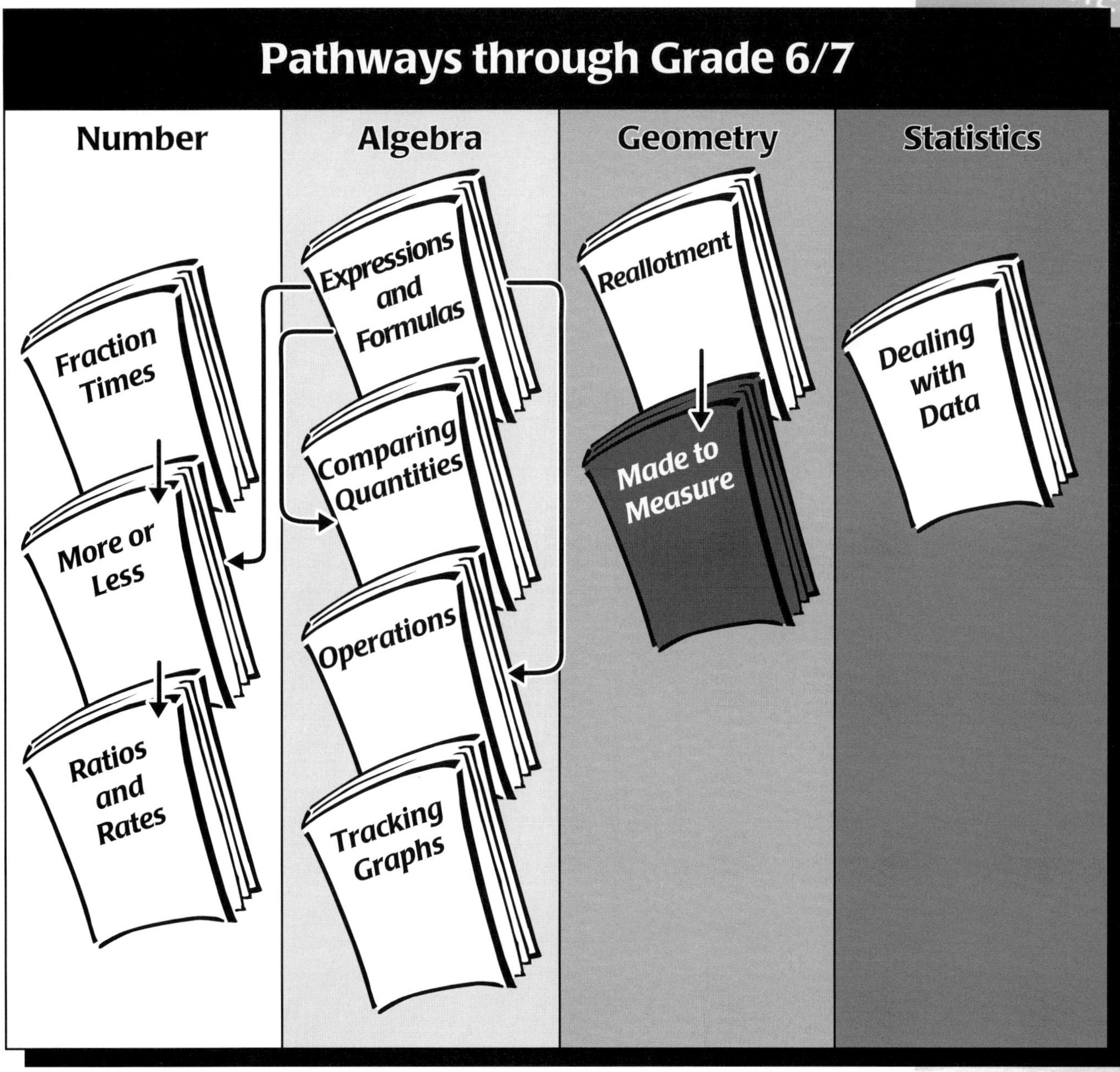

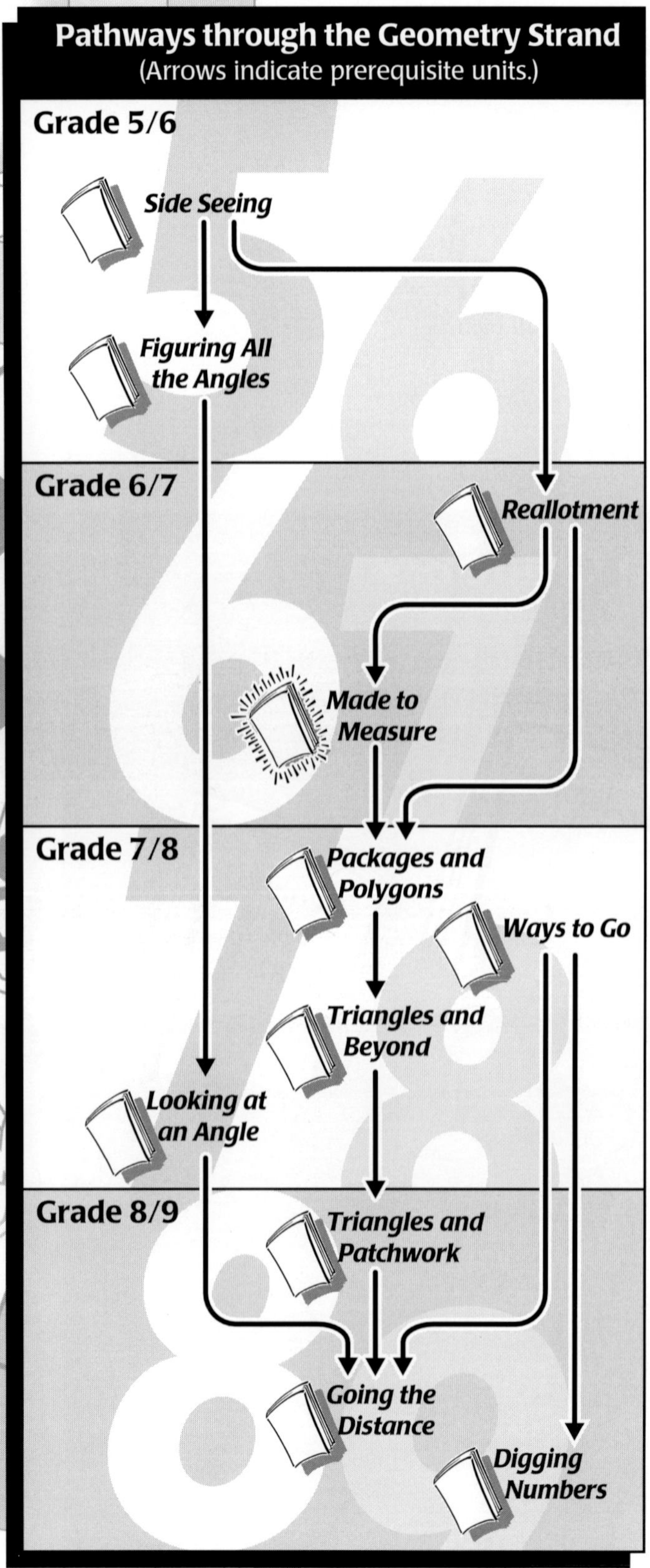

Connections within the Geometry Strand

On the left is a map of the geometry strand; this unit, *Made to Measure,* is highlighted.

Made to Measure is the fourth unit in the geometry strand and the second grade 6/7 geometry unit, following the unit *Reallotment.*

There are three central themes that run through the *Mathematics in Context* geometry strand units: orientation and navigation, shape and construction, and visualization and representation. *Made to Measure* encompasses ideas primarily from the latter two themes.

Figuring All the Angles is the first geometry unit that focuses on concepts from the shape and construction theme. In that unit, students learn to identify and measure angles and explore interior angles in various shapes. *Reallotment* is the first grade 6/7 unit that addresses ideas from the same theme. In that unit, students measure and calculate the perimeters and areas of quadrilaterals, circles, triangles, and irregular polygons. Measurement is revisited in *Made to Measure,* in which students use standard and nonstandard units to measure length, area, volume, and angles.

Both *Reallotment* and *Made to Measure* embody ideas from the visualization and representation theme. In *Reallotment,* students use visualizations and representations to find the areas of geometric figures by reshaping them and by using small, easy-to-count units. As students determine the surface area and the volume of their own bodies in *Made to Measure,* they use square and cubic models to represent the human form.

Made to Measure is the only unit dedicated exclusively to the topic of measurement. However, measurement is woven throughout all of the *Mathematics in Context* units. Therefore, it is important that students know and are able to use English and metric measurement units to estimate, measure, and compute lengths, areas, volumes, and angles and to develop a frame of reference for common measurement units to solve problems.

The Geometry Strand

Grade 5/6

Side Seeing
Exploring the relationship between three-dimensional shapes and drawings of them, seeing from different points of view, and building structures from drawings.

Figuring All the Angles
Estimating and measuring angles and investigating direction, vectors, and rectangular and polar coordinates.

Grade 7/8

Packages and Polygons
Recognizing geometric shapes in real objects and representations, constructing models, and investigating properties of regular and semi-regular polyhedra.

Looking at an Angle
Recognizing vision lines in two and three dimensions; identifying and drawing shadows and blind spots; identifying the isomorphism of vision lines, light rays, flight paths, and so forth; understanding the relationship between angles and the tangent ratio; and computing with the tangent ratio.

Ways to Go
Reading and interpreting different kinds of maps, comparing different types of distances, progressing from one two-dimensional model to another (from a diagram to a map to a photograph to a graph), and drawing graphs and networks. (*Ways to Go* is also in the statistics strand.)

Triangles and Beyond
Exploring the interrelationships of the sides and angles of triangles as well as the properties of parallel lines and quadrilaterals; constructing triangles; and using transformations to become familiar with the concepts of congruence and similarity.

Grade 6/7

Reallotment
Measuring regular and irregular areas; discovering links between area, perimeter, surface area, and volume; and using English and metric units.

Made to Measure
Measuring length (including circumference), volume, and surface area using metric units.

Grade 8/9

Triangles and Patchwork
Understanding similarity and using it to find unknown measurements for similar triangles, and developing the concept of ratio through tessellation.

Going the Distance
Using the Pythagorean theorem to investigate distances, scales, and vectors and using slope, tangent, area, square root, and contour lines.

Digging Numbers
Using the properties of height, diameter, and radius to determine whether or not various irregular shapes are similar; predicting length using graphs and formulas; exploring the relationship between three-dimensional shapes and drawings of them; and using length-to-width ratios to classify various objects. (*Digging Numbers* is also in the statistics strand.)

Connections with Other *Mathematics in Context* Units

Made to Measure is a grade 6/7 unit in the geometry strand in which students estimate and measure lengths, areas, volumes, and angles. Students were introduced to these measurement concepts in the units *Figuring All the Angles* and *Reallotment.* Angles are further studied in the grade 7/8 unit *Looking at an Angle. Made to Measure* makes connections to many other units in which measurement concepts are involved.

The following mathematical topics that are included in the unit *Made to Measure* are introduced or further developed in other *Mathematics in Context* units.

Prerequisite Topics

Topic	**Unit**	**Grade**
length	*Measure for Measure**	5/6
length, scale	*Grasping Sizes**	5/6
	Figuring All the Angles	5/6
angles	*Figuring All the Angles*	5/6
area	*Reallotment*	6/7
volume	*Reallotment*	6/7

Topics Revisited in Other Units

Topic	**Unit**	**Grade**
angles	*Looking at an Angle*	7/8
	Packages and Polygons	7/8
	Triangles and Beyond	7/8
	Triangles and Patchwork	8/9
	Going the Distance	8/9
length	*Cereal Numbers**	7/8
	Looking at an Angle	7/8
	*Growth***	8/9
	*Digging Numbers****	8/9
	*Statistics and the Environment****	7/8
	*Insights into Data****	8/9
	Going the Distance	8/9
area	*Reallotment*	6/7
	*Cereal Numbers**	7/8
	*Building Formulas***	7/8
	*Ups and Downs***	7/8
	Packages and Polygons	7/8
	Triangles and Patchwork	8/9
	*Insights into Data****	8/9
volume	*Reallotment*	6/7
	*Cereal Numbers**	7/8
	*Insights into Data****	8/9
metric system	*Powers of Ten**	7/8
arrow language	*Expressions and Formulas***	6/7

* These units in the number strand also help students make connections to ideas about geometry.
** These units in the algebra strand also help students make connections to ideas about geometry.
*** These units in the statistics strand also help students make connections to ideas about geometry.

BRITANNICA

Mathematics in Context

Student Materials and Teaching Notes

Student Book
Table of Contents

Dear Student,

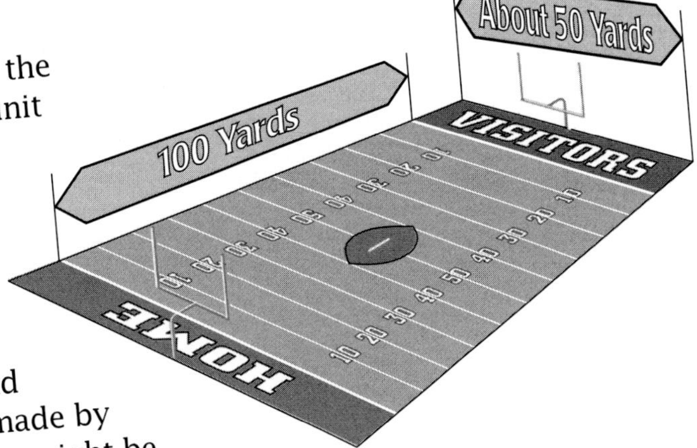

You will soon be starting the *Mathematics in Context* unit *Made to Measure.* The unit is all about measuring: measuring your feet, hands, and thumbs; measuring wood, a football field, and a deck of cards; and measuring the angles made by your arm and wrist. You might be amazed by what you can measure!

You will find that mathematics plays an important role in measurement. Every time you measure something, you might ask yourself the following questions:

- Will every person measuring this item get the same measurement that I did?
- Do all of these things have the same measurement?
- What other units of measure can I use?
- Are there other ways to measure these things?

Whenever you make a measurement in this unit, picture how big—or small or steep or short—that measurement is. When you can do this with all of the measurements in this unit, you are well on your way to becoming a mathematician!

Sincerely,

The Mathematics in Context Development Team

Work Students Do

Students begin by measuring the lengths of common objects using historic measures, such as the thumb, hand span, and fathom. They compare the class results of the various measurements and see that students' measurements for the same object varies. Students then realize the importance of having standard measures, and explore measuring objects using the metric and English systems. They investigate to see whether there is a relationship between shoe size and foot length and explore different shoe-size measurement systems used in various countries. Students also study the relationship between units of measure associated with the human body, such as foot length and pace, and learn how standard measures, such as the mile, were created. Finally, students look at alternate measures of length by using units of time to measure travel distances.

Goals

Students will:

- estimate lengths, areas, volumes, and angles using appropriate units;
- find (compute, measure, and draw) lengths, areas, volumes, and angles using standard and nonstandard units;
- use simple relationships between and within measurement systems;*
- recognize that different units of measurement and different measurement systems exist;
- recognize the importance of standard (measurement) units;*
- use formulas and rules of thumb to estimate and calculate length, area, and volume;*

* *These goals are introduced in this section and are assessed in other sections in the unit.*

Pacing

- approximately four 45-minute class sessions

Vocabulary

- fathom
- foot
- hand span
- metric system
- mile
- pace
- Scottish thumb
- yard

About the Mathematics

Students should already be familiar with both metric and English units of measure for length. Historically, many units of length measure, such as the foot, thumb, fathom, and pace, have been associated with the human body. Since measurements taken using different human body parts vary greatly, people saw the need for standard measurement units. Today, lengths are measured in either metric or English system units. The metric system is used by the majority of countries in the world. Only the United States, England, and a few other countries still use the English system.

The ability to use known lengths of body parts to estimate an object's length, width, height, surface area, and volume is a valuable skill. Students also explore the relationships between thumb length, height, and other measurements associated with the human body. This is a statistical task that typically uses a scatter plot to display the data results, as illustrated on page 21 of this Teacher Guide.

Materials

- centimeter rulers, pages 13, 15, 17, 21, 23, and 29 of the Teacher Guide (one per student)
- 5-meter piece of rope, page 13 of the Teacher Guide (one per class)
- inch rulers, pages 13, 17, 21, 23, and 29 of the Teacher Guide (one per student)
- inch or centimeter tape measures, pages 23 and 29 of the Teacher Guide (one per student)
- world map, pages 25 and 27 of the Teacher Guide, optional (one per class)
- pennies, page 29 of the Teacher Guide, optional (10 per student)

Planning Instruction

Throughout this unit, you may want students to visually estimate every measurement before they actually measure it. This will help to develop students' ability to estimate lengths, areas, volumes, and angles using appropriate metric and English measurement units. Be sensitive to students who may feel uncomfortable taking personal measurements in class. You might suggest that students take some or all such measurements as a homework assignment. It might also be useful to make and display samples of common measurement references, such as a square meter, in the classroom.

Students may work in small groups on problems 1–8, 16a, 16c, 23, 25, and 27–29. They may work on problems 11, 13–15, 17, 18, 22, 24, 26, 30, and 31 individually. Problems 9, 12, and 16b are designed as whole-class activities. Students may work on the remaining problems individually or in small groups.

Problem 21 is optional. If time is a concern, you may omit this problem or assign it as homework.

Homework

Problems 11 and 13–15 (page 14 of the Teacher Guide), 17 and 18 (page 16 of the Teacher Guide), 24 (page 20 of the Teacher Guide), 26 (page 22 of the Teacher Guide), and 30 (page 26 of the Teacher Guide) may be assigned as homework. The Extension activities (pages 9, 21, 27, and 29 of the Teacher Guide), the Bringing Math Home activities (pages 15 and 19 of the Teacher Guide), and the Writing Opportunity (page 27 of the Teacher Guide) may also be assigned as homework. After students complete Section A, you may assign appropriate activities from the Try This! section located on pages 35–37 in the *Made to Measure* Student Book as homework. The Try This! activities reinforce the main mathematical concepts covered in this section.

Planning Assessment

- Problem 31 can be used to informally assess students' ability to estimate lengths using appropriate units; find (compute, measure and draw) lengths using standard and nonstandard units; recognize that different units of measurement and different measurement systems exist; and to select and use appropriate measurement units.

A. LENGTHS

INTRODUCTION

People who design the objects you use every day have thought a lot about how big or how small those objects should be. Knowing the sizes of people's arms, legs, and hands can be very useful when designing furniture, clothes, toys, windows, doors, and many other items.

1. Which body measures would be useful to know if you were designing the following:

a. doors

b. school desks

c. shoes

d. pants

e. baby cribs

f. stairs

2. For what other objects would you need to know body measures?

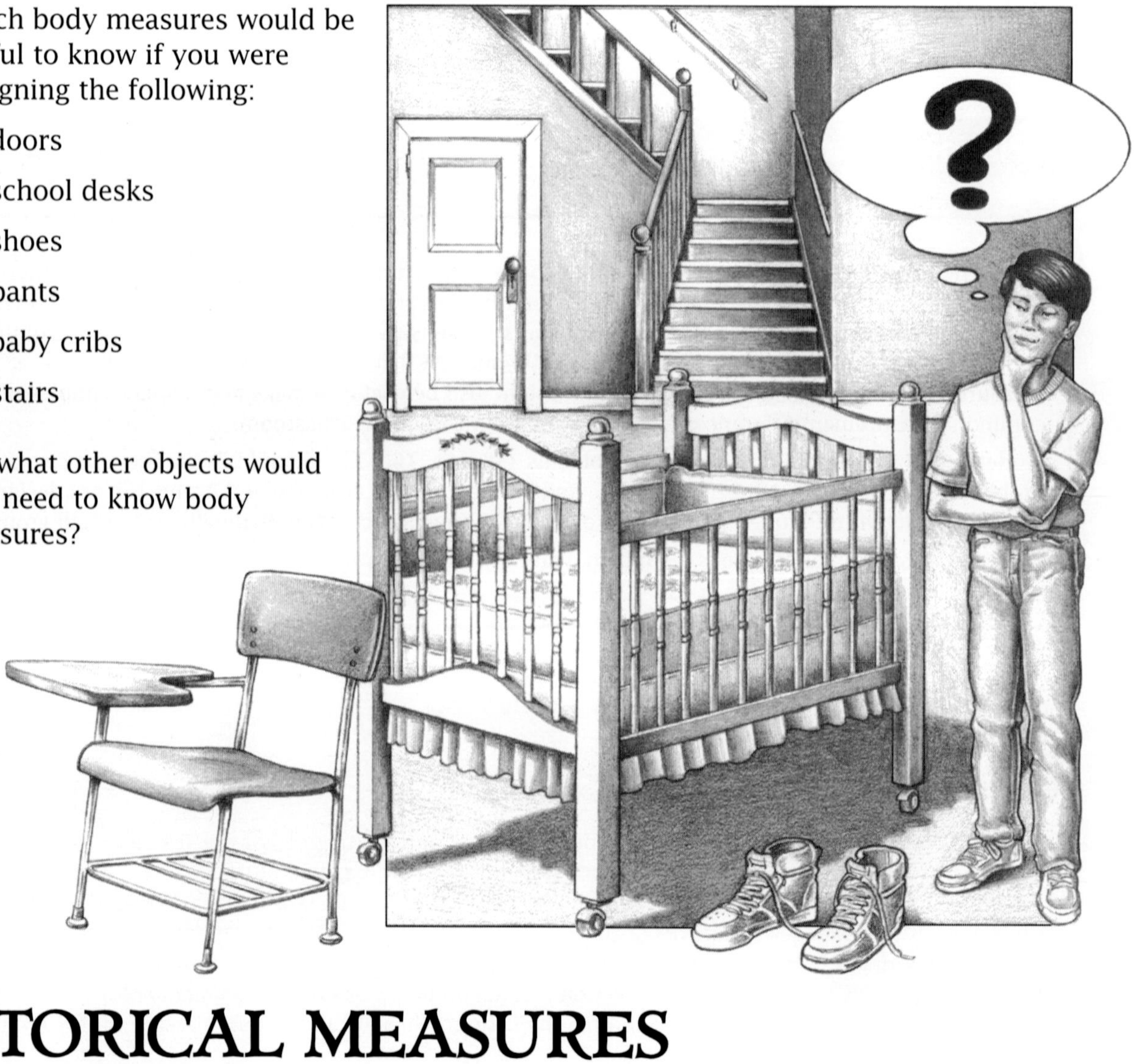

HISTORICAL MEASURES

You probably discovered that the lengths and heights of different body parts are important for designing many common objects. At one time, all measurements for length were related to the human body. Some of these units of measure include the *thumb, hand span, foot, yard, pace,* and *fathom.*

3. Match each of the units of measure listed in the above paragraph with its drawing on page 2.

Solutions and Samples
of student work

1. Answers will vary. Sample responses:
   a. Doors: height and width of person; size of a person's hand (for door knob size)
   b. School desks: sitting height; length from knee down; width of person
   c. Shoes: length and width of feet
   d. Pants: length from waist to ankle; hip size; waist size; inseam or leg length
   e. Baby cribs: height of children; width of head
   f. Stairs: length of foot

2. Answers will vary. Sample responses:
   - seats in cars
   - seats in theaters
   - hallways
   - handgrips on bicycles or exercise equipment
   - railings for stairs
   - height of bicycle seats

3. **a.** hand span
   b. yard
   c. fathom
   d. thumb
   e. foot
   f. pace

Hints and Comments

Overview Students think about body measures that might be useful to know when designing furniture, clothes, and other items. They also learn the names of some historical units of measure for length that are related to the human body and match each measure with its drawing.

Planning Students may work in small groups on problems **1–3.** Be sure to discuss students' answers for all three problems in class.

Comments about the Problems

2. Explain to students that specific measurements for certain items are required by law to ensure safety. For example, the spaces between the side bars of baby cribs must be small enough so that a baby's head cannot fit through them. Toys are sometimes required to have warning labels if they contain small parts that might be swallowed by young children. Specific measurements, such as optimal chair heights or distances from the eyes to computer terminals, are important in the workplace.

3. b. The distance from the nose to the fingertips is a common way to estimate the length of fabric material or thread.

d. The thumb is a measurement unit that refers to the width of the thumb rather than its length.

f. The pace is a measurement unit that refers to the distance from the back heel of one foot to the front toe of the other foot. The pace represents the distance covered by a typical step of a person while walking. Also see the Hints and Comments for problem **10** on page 15 of the Teacher Guide.

a.

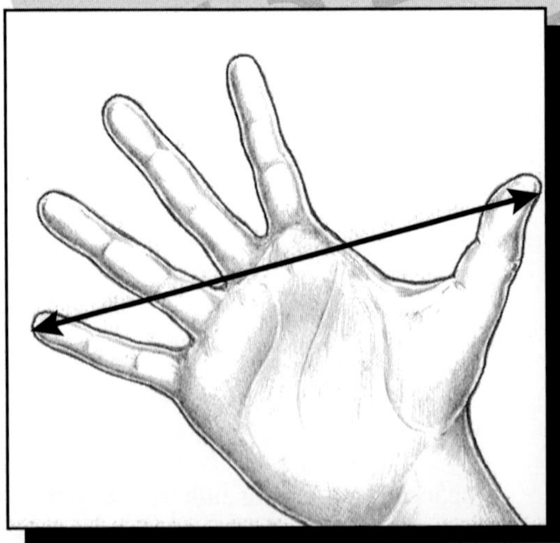

b.

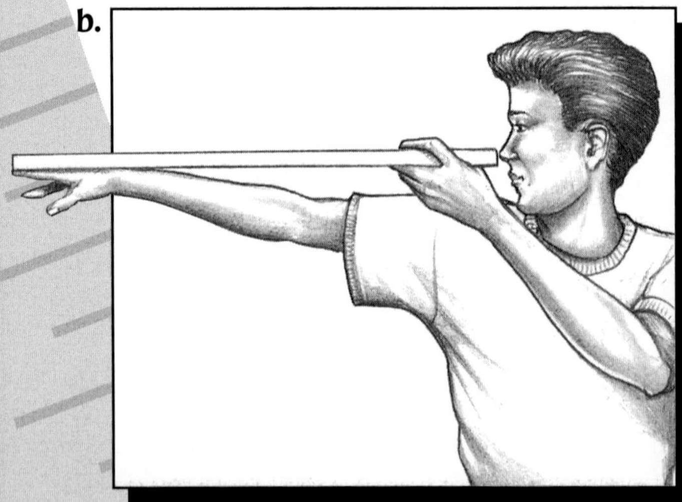

c.

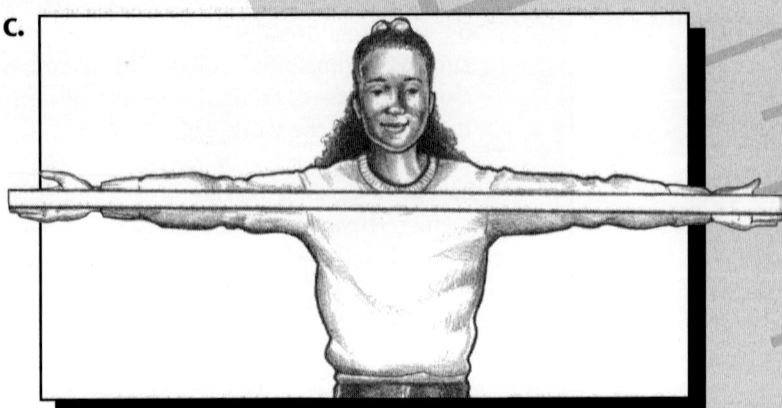

d.

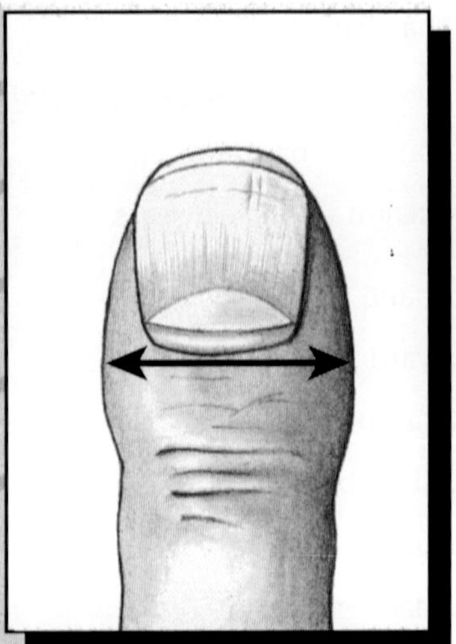

e.

f.

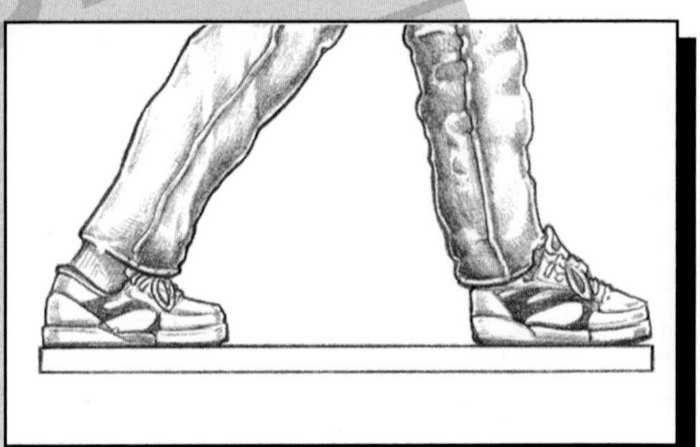

Solutions and Samples
of student work

Hints and Comments

Overview There are no problems for students to solve on this page.

Extension Have students do research about historical nonstandard units of measure. Isaac Asimov's book *Realm of Measure* is a good resource. You might also have students investigate the history of terms, such as *fathom*. The fathom is currently used to measure water depth. Have students write short reports and share their findings in class.

4. a. Which of the units of measure from problem **3** would you use to find the length of the nail shown on the right?

b. How long is this nail in the units of measure you chose for your answer to **a?**

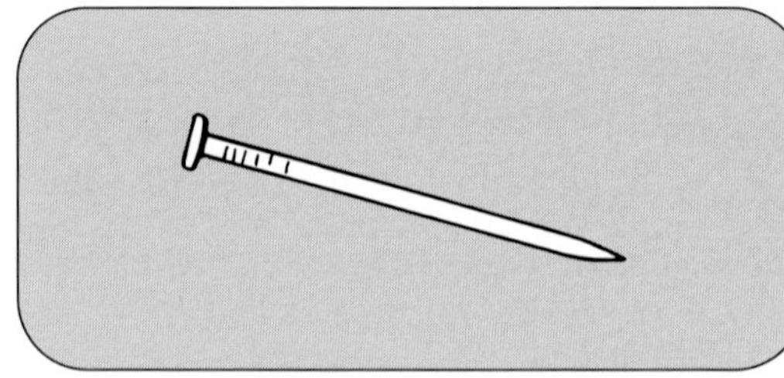

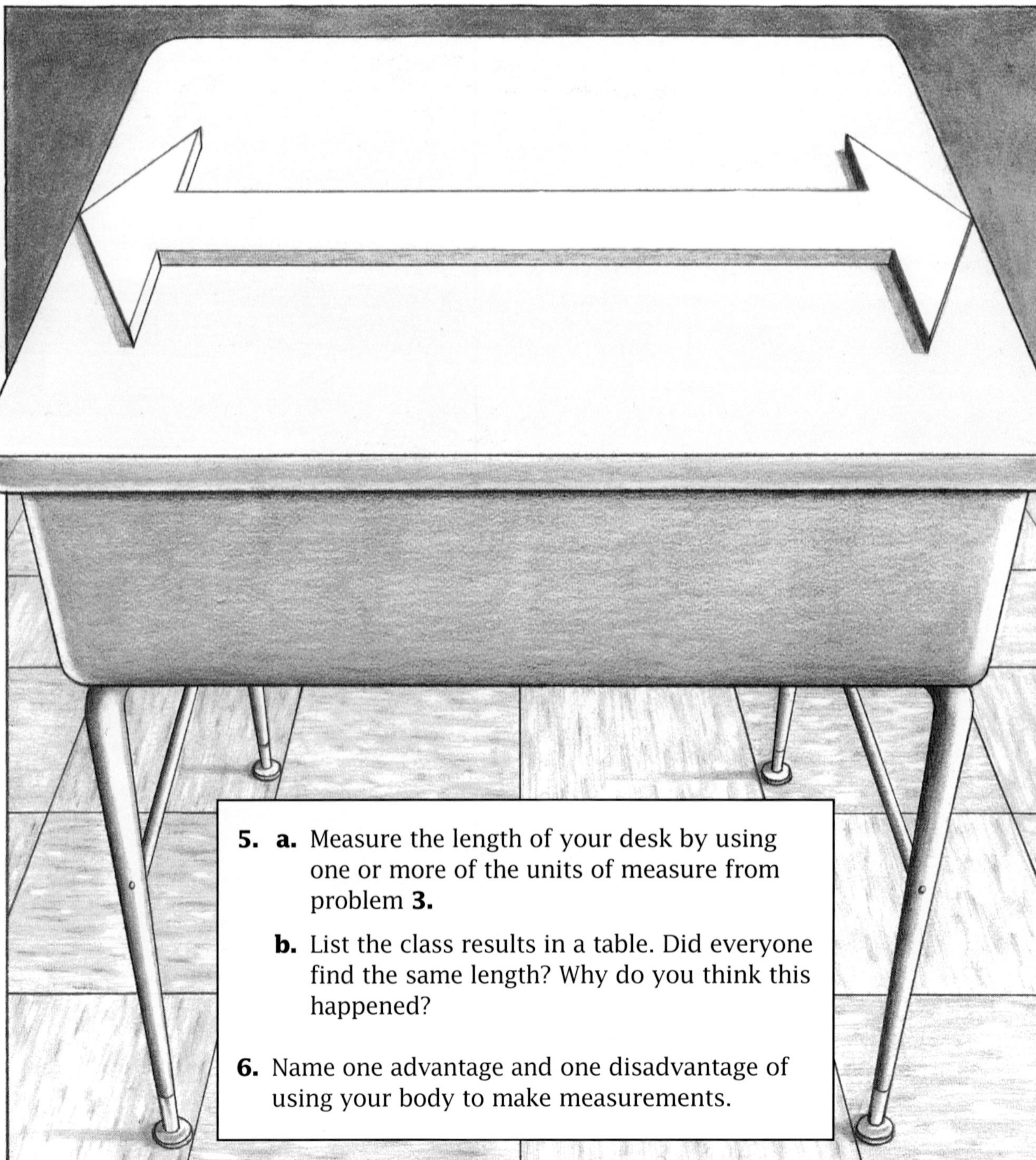

5. a. Measure the length of your desk by using one or more of the units of measure from problem **3.**

b. List the class results in a table. Did everyone find the same length? Why do you think this happened?

6. Name one advantage and one disadvantage of using your body to make measurements.

Solutions and Samples
of student work

4. a. The thumb is the most appropriate unit to measure the length of the nail.

b. The nail is about 2 or 3 thumbs long.

5. a. Answers will vary, depending on the lengths of individual desks. Accept any reasonable lengths measured in hand spans, feet, or a combination of hand spans and thumbs.

b. Students will probably have different answers even if they used the same measures because of the variation in the sizes of body parts and the level of precision used by different students.

6. Answers will vary. Sample responses:

One advantage of using your body to make measurements is that you don't need any special tools, such as rulers.

One disadvantage of using your body to make measurements is that, since different students' body parts are different sizes, our answers will not be the same.

Hints and Comments

Overview Students choose appropriate nonstandard measurement units from problem **3** to measure the lengths of a nail and their desks. They then take and compare the class results.

About the Mathematics Problems **5b** and **6** give students an opportunity to talk about using appropriate measurement units, the variability in measurement when using nonstandard units, and the possible need for standardizing units.

Measurements made using nonstandard units may result in answers that are not whole numbers. There are two methods that students might use to refine such measurements:

- use a combination of different measurement units in measuring one item, such as hand spans and thumbs, or
- use fractional parts of the measurement unit chosen. For example, the length of the desk might be $3\frac{1}{2}$ hand spans.

Planning Students may work on problems **4–6** in small groups. Discuss students' solutions and strategies for both problems.

Comments about the Problems

4. b. Note that students should take the measurement using a nonstandard unit from their own bodies, such as their own thumb, and not the units as pictured on page 2 of the Student Book.

5. b. This question can be used as the basis of a class discussion on measurement variation using nonstandard units. Possible discussion topics include: the difficulty in measuring some objects using body parts as measurement units, the rationale behind their choices of measurement units, and the difference in various body measures from one individual to the next.

6. Possible answers may arise from the class discussion in problem **5b.**

Extension As an extension of problem **6,** ask students: *How can the disadvantages of using body parts to make measurements be resolved?* [Answers may vary. Some students might suggest that the class should decide on a single foot or hand span that everyone can use, such as the average foot or hand span of the students in the class.]

In Scotland during the Middle Ages, a unit of measure called the Scottish thumb was used. A *Scottish thumb* is the mean of the thumb widths of three men: a large man, an ordinary-sized man, and a small man.

7. Why were three different-sized men used to determine the Scottish thumb?

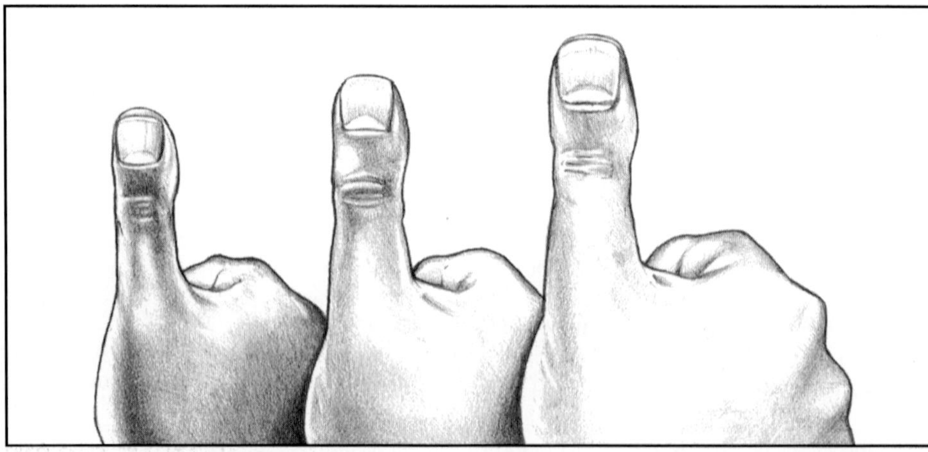

In 1616, the Germans decided to create a unit of measure called the *mean foot*. To do this, they cut a piece of rope that was as long as the feet of 16 men.

8. a. How do you think the rope was used to find the length of the mean foot?

b. Which measurement is closer to the average person's measurement: the German mean foot or the Scottish thumb? Explain your answer.

9. With the help of 16 classmates, find the length of the mean foot in your class by using the method described above.

Solutions and Samples
of student work

7. The Scotts wanted to find an average width for the thumb unit of measurement, so they used the thumbs of three different-sized men.

8. a. Since there were 16 men, the mean foot was determined by folding the rope into 16 pieces (by folding it in half four times).

b. Answers will vary. Some students might say that the mean foot is closer to the average person's measurement, since this unit was determined using the feet of 16 people, while the Scottish thumb was determined using the thumbs of only three people.

9. Answers will vary. The length of the mean foot should be expressed in centimeters or inches. In one pilot class, the total length of the feet of 16 students was about 379 centimeters, and the length of the mean foot was about 24 centimeters.

Hints and Comments

Materials centimeter or inch rulers (one per student); 5-meter piece of rope (one per class)

Overview Students learn the origin and meaning of a unit of measure called the Scottish thumb. They also read about how the Germans used 16 men to determine the length of the mean foot. Students find the mean foot of 16 classmates using the same method.

About the Mathematics The concept of using a representative sample is implied in the methods used by both the Scotts and the Germans to establish the two nonstandard units presented on Student Book page 4. In the unit *Dealing with Data,* students learned that a representative sample is a sample that has characteristics similar to the population and in roughly the same proportion as that of the population.

Planning Students may work on problems **7** and **8** in small groups. Be sure to discuss students' answers to problem **8.** You might want to do problem **9** as a whole-class activity.

Comments about the Problems

8. a. Notice that the Germans chose the length of the feet of 16 men to determine the mean foot. Since the number 16 is a power of two, a rope made up of the lengths of an even number of feet could be easily folded.

b. This question is intended to be open-ended to encourage students to discuss its implications. Some students may say that the German mean foot is closer to the average person's measurements since it is based on the foot lengths of 16 men. Other students might prefer using the Scottish thumb since it is easier to find. Students may also comment on the variability of the two nonstandard measures. It is likely that there will be greater differences in the lengths of feet than in the widths of thumbs.

9. If *Dealing with Data* was taught prior to this unit, you might discuss the importance of using a representative sample of 16 students. After completing this problem, students can investigate whether one student's foot is the same length as the mean foot of the class.

10. Measure the following in centimeters. List the results in a table.

- **a.** your thumb width
- **b.** your hand span
- **c.** your foot length
- **d.** your pace

11. Use the table you made in problem **10** to find the following relationships between measures:

- **a.** How many thumb widths are in one hand span?
- **b.** How many thumb widths are in one foot?
- **c.** How many feet are in one pace?

You may add other units of measure and the relationships between them to your list.

12. What is the size of the "typical" thumb for your classmates? Explain your answer.

Imagine what it must have been like when everyone used their own thumbs for measuring. Today, of course, we have standard systems of measurement.

A few countries, including the United States, still use the foot as a unit of measure, but the length of a foot no longer refers to the length of each person's foot. A standard has been officially established for the length of a foot. Most countries use the *metric system*, which was adopted in France in 1795.

13. **a.** The foot is a part of the English system of measurement. List some other units of measure for length that are part of the English system.

b. In your notebook, write as many relationships between the units of measure in the English system as you can.

14. **a.** List some units of measure for length that are part of the metric system.

b. Write as many relationships between these units of measure as you can.

15. Find some simple relationships between the English and metric systems.

Solutions and Samples
of student work

10. Answers will vary. Sample responses:

a. 2 centimeters

b. 19.5 centimeters

c. 24 centimeters

d. 58 centimeters

11. Answers will vary, depending on students' measurements for problem **10.** Sample responses based on sample answers in problem **10:**

a. Almost 10 thumbs are equal to a hand span.

b. Twelve thumbs are equal to one foot length.

c. About $2\frac{1}{2}$ foot lengths are equal to one pace.

12. Answers will vary. Some students might talk about the thumb width measurement that occurred most often (the mode), while other students may refer to the average thumb width of the class (the mean). Others may mention the middle value of the data (the median).

13. a. Common units of measure for length from the English measurement system are the inch, foot, yard, and mile.

b. Answers will vary, depending on students' knowledge of the relationships between the units. Sample responses:

12 inches = 1 foot; 36 inches = 1 yard; 3 feet = 1 yard; 5,280 feet = 1 mile; 1,760 yards = 1 mile

14. a. Common units of measure for length from the metric measurement system are the millimeter, centimeter, decimeter, meter, and kilometer.

b. Answers will vary, depending on students' knowledge of the relationships between the units. Sample responses:

10 millimeters = 1 centimeter; 10 centimeters = 1 decimeter; 100 centimeters = 1 meter; 1,000 meters = 1 kilometer

15. Answers will vary, depending on students' knowledge of the relationships between the units in both measurement systems. Sample responses:

1 meter is about 3 feet; 1 inch is exactly 2.54 cm; and 1 kilometer is about $\frac{2}{3}$ mile.

Hints and Comments

Materials centimeter rulers (one per student)

Overview Students measure their thumb width and other nonstandard measurement units in centimeters and make a table showing the results. They then use the table to describe relationships between the various nonstandard units.

About the Mathematics The metric system is based on powers of 10, which facilitates the process of converting between different metric measurement units. For example, 10 millimeters = 1 centimeter; 10 centimeters = 1 decimeter, and so on.

Planning Students may work on problem **10** individually or in small groups. Discuss problem **12** as a whole class. You may want students to work on problems **11** and **13–15** individually and assign these problems as homework.

Comments about the Problems

10. Some students may need to be reminded that the markings on rulers do not always begin at the far left edge of the ruler. Since different methods can be used to measure one's pace, students' answers may vary widely. You might wish to discuss this in class.

11. Homework This problem may be assigned as homework

13–15. Homework These problems may be assigned as homework. Be sure to discuss students' responses to evaluate their present understanding of various English and metric units and the relationships between the units within each system.

15. You may want to give some common reference points for students to use when discussing metric measurements. For example, a door is about 2 meters high, and a thumb is about 2 centimeters wide.

Bringing Math Home Have students discuss with their families and/or do additional research on English and metric measurement units with which they are unfamiliar.

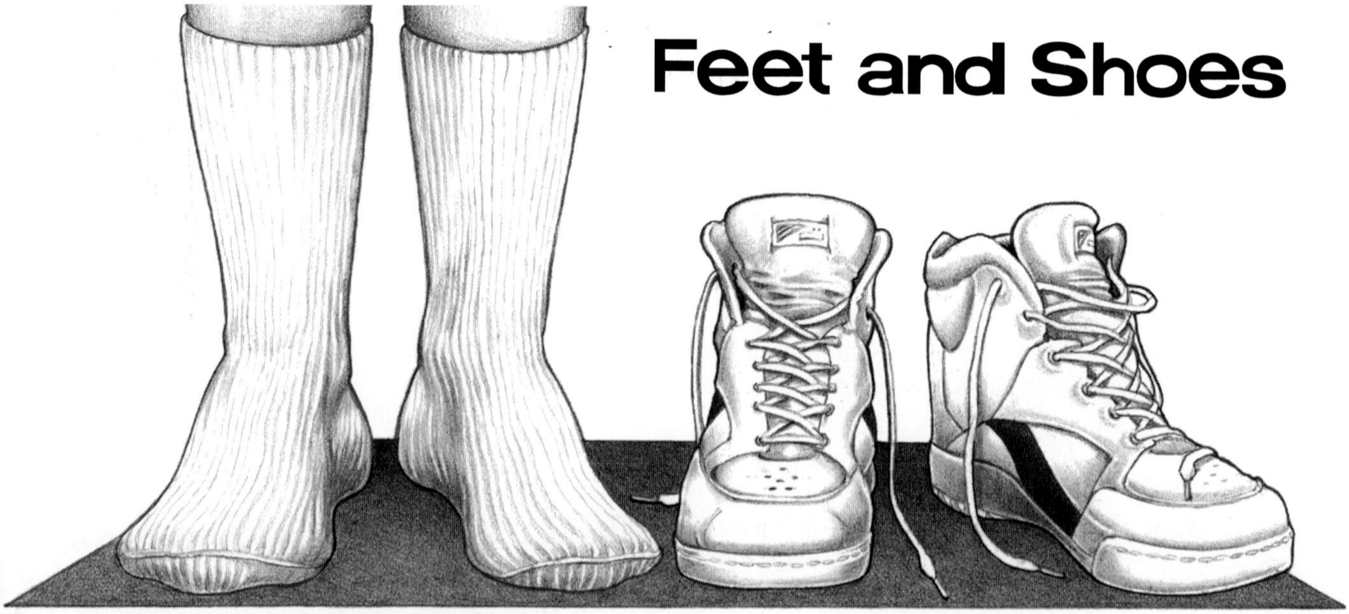

Feet and Shoes

In problem **10c,** you measured the length of your foot in centimeters. The length of your foot is different from your shoe size.

16. a. Do you think that there is a relationship between shoe size and foot length? Explain your answer.

b. Make a table that lists the foot length (in centimeters) and corresponding shoe size for each student in your class.

c. Graph the results. Put foot length in centimeters on the horizontal axis and shoe size on the vertical axis. What does your graph tell you about the relationship between foot length and shoe size?

Just as countries use different systems of measurement, they also have different systems for determining shoe sizes. For some shoes, you can find at least three different sizes:

- European size—usually a number between 33 and 47
- U.K. (United Kingdom), or English, size—usually a number between 1 and 15
- U.S. size—usually a number between 1 and 15 (slightly larger than U.K. sizes)

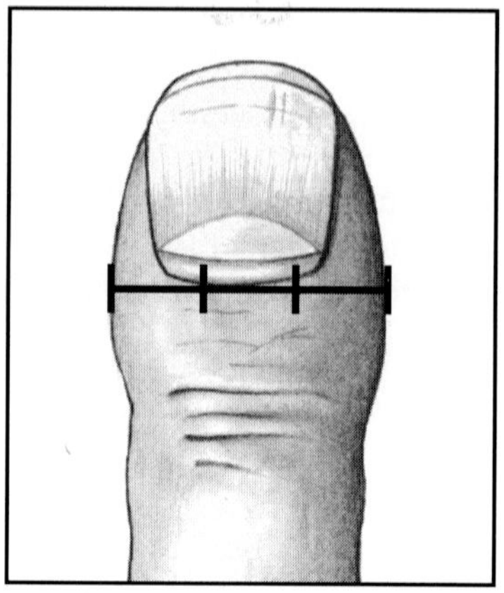

The U.K. system of shoe sizes began in the seventh century. Shoe sizes were measured with a standard thumb (now called an inch).

17. How many standard thumbs (or inches) long is your foot?

To get a more accurate measurement, the U.K. introduced a smaller unit of measure, the stitch. Three *stitches* are in one standard thumb.

18. How many stitches long is your foot?

16. a. Yes, there is a relationship between shoe size and foot length. Explanations will vary. Sample explanation:

If there were no relationship between foot length and shoe size, there would be no way for people to buy shoes that fit them without trying on many pairs of shoes.

b. Tables will vary. Sample table:

Length of Foot (centimeters)	Shoe Size	Length of Foot (centimeters)	Shoe Size
24	7.5	25	8
24	7.5	21	5.5
22	7	22	5.5
22	6.5	22	5
25	8.5	21	5
25	8	22	6
23.5	6.5	24	8
25	8	26	9.5
24	7	26	9.5

c. Graphs will vary, depending on students' data for part **b.** Sample graph based on sample table in part **b:**

The graph shows that there is a relationship between shoe size and foot length. As foot length increases, shoe size increases. The points on students' graphs should roughly form a straight line or a long oval.

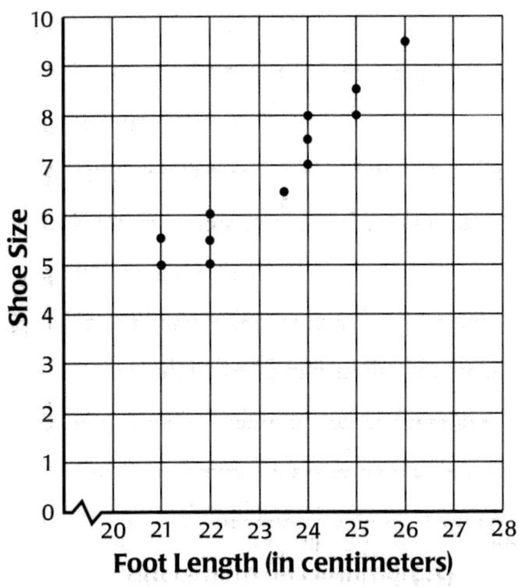

17. Answers will vary, depending on the lengths of students' feet. Sample response:

If your foot length is about $24\frac{1}{2}$ cm, then it is about $9\frac{2}{3}$ inches or 29 stitches long.

18. Answers will vary, depending on students' responses for problem **17.** Answers should be about three times the answer to problem **17.**

Materials centimeter and inch rulers (one of each per student)

Overview Students investigate the relationship between shoe size and foot length by organizing the information into a table that lists the foot lengths and corresponding shoe sizes for each student and displaying the results in a graph. They also learn about three different measurement systems for determining shoe sizes.

Planning Problems **16a** and **c** may be done in small groups. Problem **16b** should be done as a whole-class activity. You may want to have students work on problems **17** and **18** individually as a homework assignment.

Comments about the Problems

16. Since shoe-size systems differ for boys and girls, you may want to have the boys make a table and graph of their data and the girls do likewise.

c. Be sure that students understand how to construct the graph and label its axes correctly. If all students' shoes are labeled within the same shoe-size system, it is likely that students will see a relationship between shoe size and foot length.

17. Students might compare the standard thumb (one inch) to their own thumbs. The widths of students' thumbs will probably measure less than one inch, unless their thumbs are "squashed" flat. Students should use an English ruler to measure their feet since they need to find the length in thumbs (inches).

17–18. Homework These problems may be assigned as homework.

In the U.K. system, shoe sizes are based on the number of stitches. The first 25 stitches are not counted in adult shoe sizes. Size 1 is, therefore, really 26 stitches, or $8\frac{2}{3}$ inches.

19. a. Copy the table below into your notebook and continue it to shoe size 8.

Foot Length (in stitches)	Foot Length (in inches)	United Kingdom Shoe Size
26	$8\frac{2}{3}$	1
27		

b. Use your answer to problem **18** to find your U.K. shoe size in the table you made in part **a.** How does your U.K. shoe size compare with your U.S. shoe size?

c. Formulate a rule that helps you find someone's foot length in inches if you know his or her U.K. shoe size. Write the rule in arrow language.

$$\text{U.K. shoe size} \xrightarrow{?} \_\_\_ \xrightarrow{?} \text{foot length in inches}$$

20. After you finish problem **19,** look back at your answers to problem **16.** Would you change your answers now? Why or why not?

19. a.

Foot Length (in stitches)	Foot Length (in inches)	U. K. Shoe Size
26	$8\frac{2}{3}$	1
27	9	2
28	$9\frac{1}{3}$	3
29	$9\frac{2}{3}$	4
30	10	5
31	$10\frac{1}{3}$	6
32	$10\frac{2}{3}$	7
33	11	8

Note: Three stitches is about 1 inch.

b. Answers will vary. In general, U.S. shoe sizes are larger than U.K. sizes. Using the foot length of 29 stitches ($9\frac{2}{3}$ inches), the U.K. shoe size is 4.

c. U.K. size $\xrightarrow{+25}$ Stitches $\xrightarrow{\div 3}$ foot length in inches

20. Answers will vary. If students indicated in their answer for problem **16c** that there is no relationship between foot length and shoe size, they should see the need to change their answer after completing problem **19.**

Overview Students make a table of foot lengths and U.K. shoe sizes, and formulate a rule to find someone's foot length if they know the U.K. shoe size. They also compare U.K. shoe sizes to U.S. shoe sizes.

Planning Students may work on problems **19** and **20** individually or in small groups. Discuss their solutions to both problems in class.

Comments about the Problems

19. b. You might suggest that students add two columns in the table to include the U.S. shoe sizes for both men and women. Be sure that students use adult shoe sizes here rather than children's shoe sizes.

c. Students are introduced to using arrow language to express rules or formulas in the grade 6/7 unit *Expressions and Formulas.*

20. This problem asks students to find a relationship between U.K. shoe sizes and foot length in inches. In problem **16c,** students made a table and graph to investigate the relationship between foot length and their own shoe size.

Bringing Math Home Have students measure the foot lengths of family members in stitches or inches to determine their U.K. shoe size using the table they made in problem **19a** or applying the rule they formulated in problem **19c.** They can also compare the U.K. shoe sizes they found for family members with their U.S. shoe sizes.

Feet and Miles

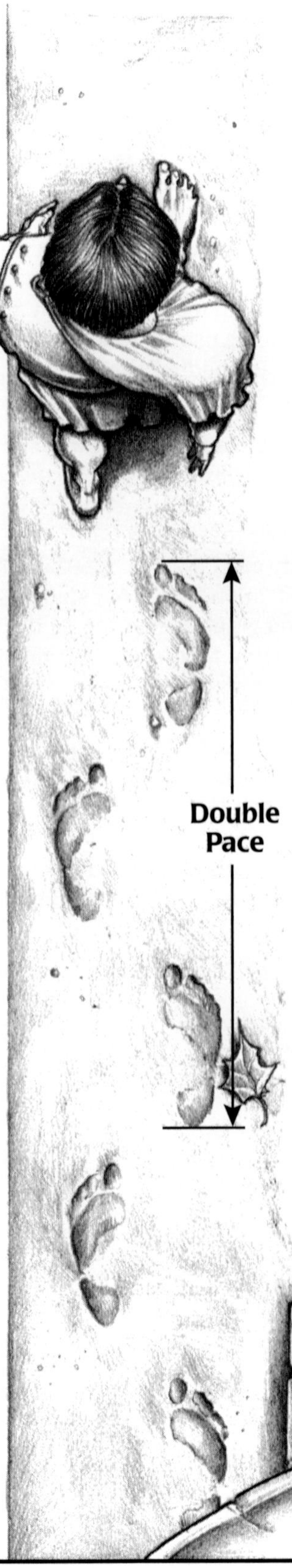

In problem **10** you measured your pace and your foot length.

21. Make a table of the pace and foot-length measurements of all of the students in your class. Use the results to determine whether the two measures are related. For example, you can find out if the person with the biggest foot also has the longest pace or if the person with the shortest foot has the shortest pace. Drawing a graph may be helpful.

The Romans invented the mile as a unit to measure distances. The word *mile* comes from *millia passuum,* which means "1,000 double paces." *Note:* The *double pace* is not two paces long. It is measured from the back of one foot to the front of the same foot after two steps have been taken.

22. Using the Roman definition of a mile being equal to 1,000 double paces, find the length of your own "personal" mile.

In 1592, a British law about a different kind of mile read:

"The statute mile is 8 furlongs, of 40 rods each, of 16.5 feet each."

A *furlong* is another old measure. The word comes from the phrase "a furrow long."

23. What is the meaning of the word "furrow"? Find the length of one furrow by using the above law.

24. **a.** Compare your personal mile with the Roman mile (which is 1,478.7 meters long).

b. Compare your personal mile with the statute mile, which is the one normally used today.

c. Write a paragraph about your results.

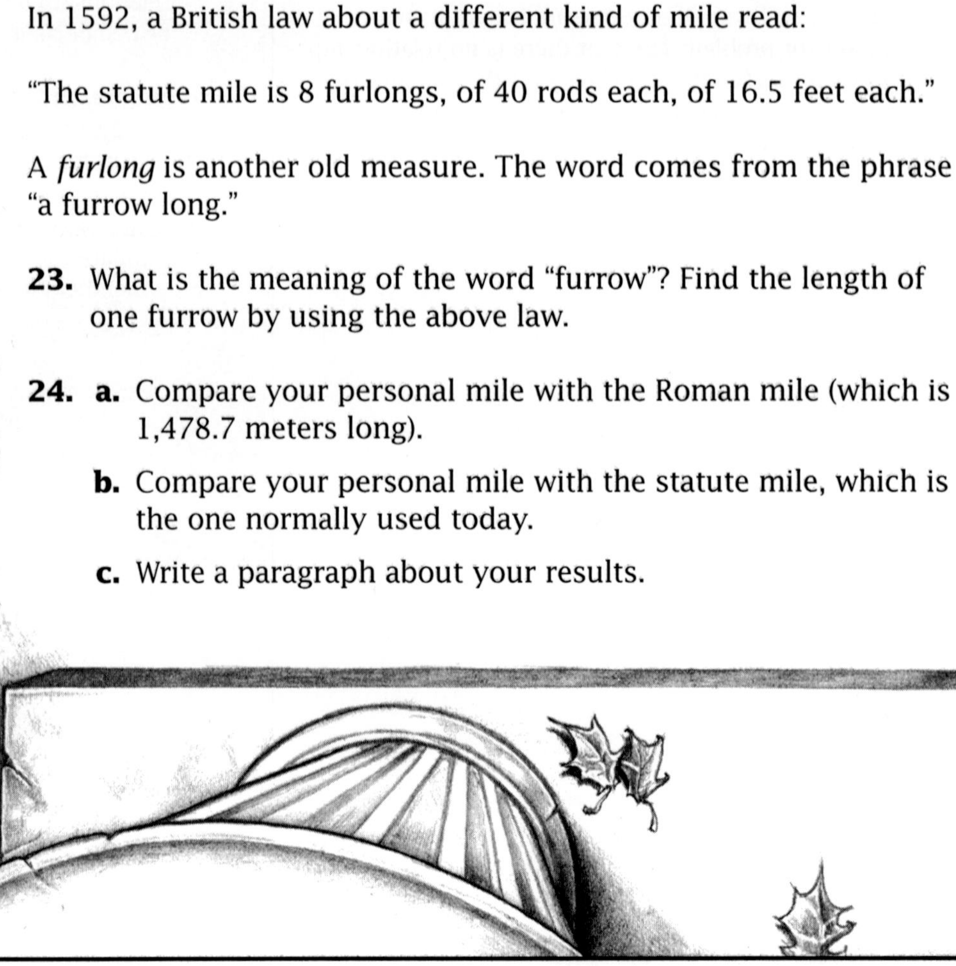

Solutions and Samples
of student work

21. Tables and graphs will vary. Students' responses should indicate that there is probably a relationship between foot length and pace. Some students may draw a scatter plot, as shown below:

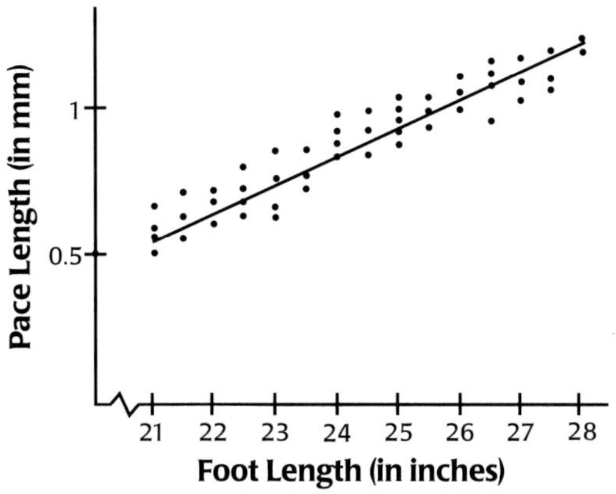

In the graph above, the relationship between pace length and foot length is about 3:1 (the length of the pace is about three times the length of the foot). Since the length of the pace is also related to the length of the legs, the above relationship is no more than a rule of thumb.

22. Answers will vary. For example, if a student's double pace were 60 inches (152.4 centimeters) long, then his or her personal mile would be 60,000 inches, 152,400 centimeters, or 1,524 meters.

23. A furrow is a trench in the earth made by a plow. The length of one furrow is 660 feet. Strategies will vary. Sample strategy:

One furlong is 40 rods; each rod is 16.5 feet; so 40 rods × 16.5 feet per rod = 660 feet.

24. a. Answers will vary, depending on students' answers for problem **22.** A personal mile of 1,524 meters is slightly larger than the Roman mile (1,478.7 meters).

b. Answers will vary, depending on students' answers for problem **22.** A personal mile of 160,000 inches is 5,000 feet which is slightly less than the statute mile. The statute mile is 8 furlongs, which is 8 × 660 feet, or 5,280 feet, or 1,609 meters.

c. Answers will vary. Sample response:

The statute mile is the longest distance at 5,280 feet (1,609 m). One's personal mile may vary but will most likely be longer than the Roman mile. The shortest distance is the Roman mile at 1,478.7 meters.

Hints and Comments

Materials centimeter or inch rulers (one per student)

Overview Students investigate the relationship between pace and foot length. They learn about the Roman mile and the statute mile, which is based on furlongs, and compare these with their "personal" mile.

About the Mathematics The length of a double pace is *not* the same as doubling the length of a pace. As in the illustration on Student Book page 2, a single pace is measured from the back of one foot to the front of the other foot. A *double* pace is measured from the back of one foot to the front of the same foot after two steps have been taken. The length of a double pace would then be equal to twice the length of a pace minus the length of one foot. You might have students measure the length of their double pace.

Planning Problem **21** is optional. You might want students to work on problem **22** individually and then complete problem **23** in small groups. Discuss students' answers in class. Problem **24** may be assigned for homework. Problem **24c** may be used as a writing opportunity.

Comments about the Problems

21. This problem challenges students to find the relationship between pace and foot length. More advanced students may use sophisticated mathematical arguments. If comparing height to shoe length will offend any of your students, you may provide data to discuss instead.

22. You might discuss with students why the double pace is not the same distance as twice the length of one pace. See the About the Mathematics section above.

23. Students may look up the meaning of the word *furrow* in a dictionary or encyclopedia. Note that since one statute mile is 8 furlongs, a statute mile is 8 × 660 feet, or 5,280 feet, our standard mile.

24. Homework This problem may be assigned as homework.

Extension You might have students complete the problems on this page using metric units.

Body Length and Fathom

The fathom is another unit of measure associated with the human body. The *fathom* originated as the distance from the middle fingertip of one hand to the middle fingertip of the other hand of a person holding his or her arms fully extended sidewards.

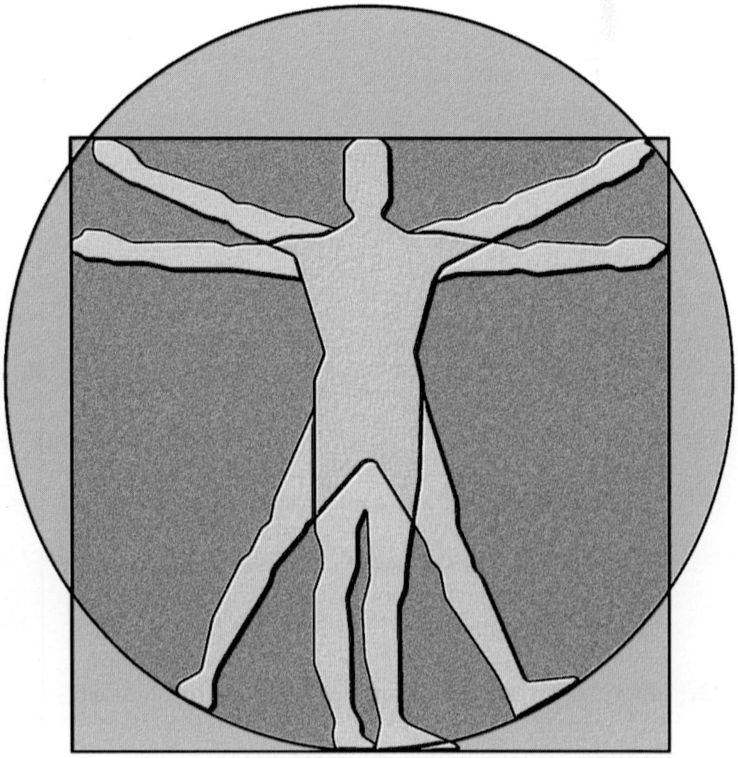

The picture of a man shown on the right is based on a famous drawing by Leonardo da Vinci. As you can see, the man more or less fits in a square.

25. Based on the pictures on this page, what is the relationship between a person's height and his or her fathom?

26. Measure your height and your fathom to decide how precisely you would fit into a square.

25. A person's height is approximately equal to his or her arm span (or fathom).

26. Most students will respond that they would fit or almost fit in a square: their arm span (or fathom) will most likely be the same length as their height.

Materials centimeter or inch rulers (one per student); inch or centimeter tape measures (one per student)

Overview Students learn the meaning of another measurement unit associated with the human body known as a fathom. They investigate the relationship between a person's height and his or her fathom and measure their own heights and fathoms to decide how precisely they would fit into a square.

Planning Problem **25** may be done in small groups. You might want students to work on problem **26** individually and assign this problem as homework.

Comments about the Problems

25. Have students use centimeter or inch rulers to measure the length between the fingertips of the horizontally outstretched arms of the person in the drawing and compare that length with the person's height. Since the person more or less fits into a square, most students will see that a person's height is approximately equal to his or her fathom.

26. **Homework** This problem may be assigned as homework. Students may measure their heights and fathoms using a metric or English tape measure.

Did You Know? Leonardo da Vinci was an Italian artist and scientist who lived from 1452–1519. The Mona Lisa was one of his famous paintings. He was also an accomplished sculptor and architect. His scientific endeavors included studying the anatomy of the human body and investigating flight.

Extension The following activity is taken from the book *Mathematics, Measurement and Me* by Doug M. Clarke. Read the following description aloud to students:

People tend to be divided into three categories: tall doors (where their height is greater than their arm span); wide doors (where their height is less than their arm span); and square doors (where their height is very close to their arm span). Ask students: *Are you a tall door, a wide door, or a square door? Explain.* [Answers will vary.] Use fractions to record the part of the class that falls into each category. Ask students: *What do the fractions total?* [1]

Other Measures for Length

There are many other ways to measure length. In Papua, New Guinea, for example, a local unit of distance is "a day's travel."

27. Why might a day's travel make sense as a unit of distance?

In mountainous regions in New Guinea, walking distances are expressed in hours, not in kilometers or miles. For example, a citizen of New Guinea might say that the distance from the village to the lake in the mountains is two hours, but the distance back is five quarters of an hour.

28. **a.** Why do you think there are two travel times?

b. Is the distance in kilometers different for the two directions? Explain.

Solutions and Samples
of student work

27. Answers will vary. Sample response:

If a person is traveling in the mountainous areas of Papua, New Guinea, the time it takes to travel across the terrain on narrow and winding roads would be more useful than knowing the distance between two given points. A person might need to determine when to leave home in order to arrive at another point at a certain time.

28. **a.** Since it requires more time to walk up the mountain than it does to walk down the mountain, there are two different travel times.

b. The distance remains the same if the same path is used on both trips.

Hints and Comments

Materials world map, optional (one per class)

Overview Students learn about other ways to measure length, such as the distance a person can travel in one day. Students reason about how such a unit of distance makes sense. They also compare this distance unit with distance measured in kilometers.

About the Mathematics The concept of using distance units other than length is not new for most students. In the grade 5/6 unit *Figuring All the Angles,* students are introduced to the difference between as-the-crow-flies distance and "taxicab distance."

Planning Students may work on problems **27** and **28** in small groups. Be sure to have a class discussion after students complete the problems on this page.

Comments about the Problems

27. You may want to locate Papua, New Guinea, an island in the Pacific Ocean north of Australia, on a world map and discuss its mountainous terrain.

28. You might discuss how distance is figured in mountainous areas, such as by travel time or as-the-crow-flies distance. Actual distances in these areas are difficult to measure. The distances between two points are not often given as precise road distances.

In the 14th century, the biggest trading empire in Africa was the Empire of Mali. Mansa Musa was one of its emperors. Sheik Uthman ed-Dukkali, a learned Egyptian who lived in Mali for 35 years, declared that Mali was "four months of travel long and four months wide."

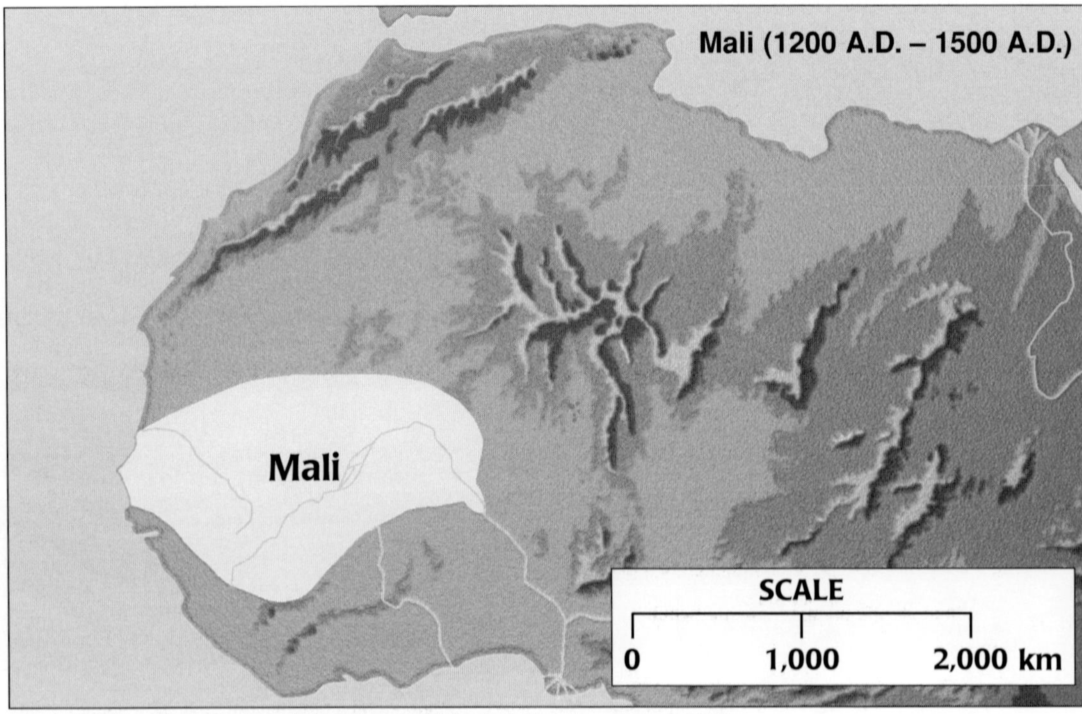

Source: Data from Basil Davidson, *African Kingdoms* (New York: Time Incorporated, 1966).

29. a. Use the above picture to estimate the length and width of the Empire of Mali in both English and metric units of measure. (*Note:* 500 miles = 800 kilometers)

b. What is the distance in miles or kilometers of "one month of travel?"

c. Based on your answer to part **b,** how do you think people in Mali traveled in the 14th century?

30. List some distances that are usually expressed in units of time rather than length. Explain why using time is appropriate in each case.

Solutions and Samples

of student work

29. a. The Empire of Mali measures about 2,250 kilometers (length) by 1,200 kilometers (width).

b. about 18 kilometers or 10 miles per day. Explanations will vary. Sample explanation:

2,250 kilometers in 4 months is equal to about 550 kilometers per month (2,250 ÷ 4 = 562.5), which is equal to about 18 kilometers per day (550 ÷ 30 = 18.5), or about 10 miles per day.

c. Answers will vary. The people of Mali could have traveled a distance of 18 kilometers per day on foot, which is a 4–5 hour trip, depending on the type of terrain.

30. Answers will vary. Sample responses:

- the distance that students live from school, a shopping mall, or the library.
- the distance from the city in which students live to a larger city in the area or to a different state.

Hints and Comments

Materials a world map, optional (one per class)

Overview Students use a map scale and information about the size of the Empire of Mali expressed in travel time to investigate the distance in kilometers that people could have traveled in one month.

About the Mathematics Students further investigate the differences between maps indicating road distances and maps based on travel times in the unit *Ways to Go.*

Planning You might want students to work on problem **29** in small groups. Problem **30** may be assigned as homework. Discuss students' answers for both problems during the next math class.

Comments about the Problems

29. a. You might want to locate Africa on a world map to examine the current boundaries of the country of Mali today. You could also compare the relative size of the empire (and/or country) of Mali with that of the United States.

c. This question is intended to help students get a sense of how fast people traversed Mali carrying cargo many years ago. Students with some previous mountain hiking experience might recall that locations are often referred to by how many days of travel they are from the base of the mountain.

30. Homework This problem may be assigned as homework.

Extension Have students ask several people how far they live from home to some location, such as their place of work or a shopping mall. Students can note whether people respond using units of distance or time and share their findings with the class.

Writing Opportunity Have students write a paragraph about the Mali Empire, which flourished in western Africa from about 1240 to 1500 A.D., and the contributions of Mansa Musa, who reigned as emperor from 1312 to 1337 A.D. The empire stretched from the northern coast of modern Senegal along the Atlantic Ocean to modern Nigeria and from the rain forests in the south to oases in the Sahara Desert.

Summary

In this section, you learned many different ways to estimate and measure length. In the past, people used thumbs, feet, and arms to measure length, but the same distance was often measured differently by each person.

Today, two standard systems of measurement exist. Most countries of the world use the metric system, in which length is measured in centimeters, meters, and kilometers. A few countries use the English system, in which length is measured in inches, feet, yards, and miles.

Summary Questions

31. Measure the following lengths any way you like. Discuss your answers in class.

a. the distance from your home to school

b. the length of your classroom

c. the thickness of a penny

Solutions and Samples
of student work

31. a. Answers will vary. Students might express the answer in distance units, such as paces, meters, or miles, or in time units.

b. Answers will vary. Many classrooms are about 8 meters, or about 25 feet long.

c. A penny has a thickness of about 1.5 millimeters.

Hints and Comments

Materials metric or English tape measures, optional (one per student); inch or centimeter rulers, optional (one per student); pennies, optional (10 per student)

Overview Students read and discuss the Summary that reviews the nonstandard measurement units associated with the human body. They then measure three lengths using any measurement unit and discuss their findings in class.

Planning You may want students to work individually on problem **31.** This problem may be used as informal assessment. Discuss students' solutions and strategies in class.

Comments about the Problems

31. Informal Assessment This problem assesses students' ability to estimate lengths using appropriate units; to find (compute, measure, and draw) lengths using standard and nonstandard units; to recognize that different units of measurement and different measurement systems exist; and to select and use appropriate measurement units.

c. Encourage students to measure a stack of pennies and then divide the total height of the stack by the number of pennies to determine the thickness of one penny.

Extension Have students select more objects or distances and choose appropriate units of measurement to determine their lengths.

Work Students Do

In this section, students use different estimation methods to approximate their own surface areas. They first estimate their body surface area by making a life-size version of a square (formed by a student's height and armspan) out of newspaper. They shade and find the area of three-fifths of the newspaper square in square centimeters, which approximates their surface area. Students also use the rule of thumb that about 100 hand prints cover the body to estimate their own surface area. They trace their hand on grid paper and estimate the area of their handprint in square centimeters. Students then use the "rule of nines" to calculate their surface area. The rule of nines divides the body into 11 sections, each of which accounts for 9% of the total surface area. At the end of this section, students learn about a nomogram chart that can be used to determine body surface area if a person's height and weight are known.

Goals

Students will:

- estimate lengths, areas, volumes, and angles using appropriate units;*
- find (compute, measure, and draw) lengths, areas, volumes, and angles using standard and nonstandard units;*
- use simple relationships between and within measurement systems;*
- recognize that different units of measurement and different measurement systems exist;
- recognize the importance of standard measurement units;
- use formulas and rules of thumb to estimate and calculate length, area, and volume;
- develop and use a frame of reference for measurement to solve problems;
- select and use appropriate measurement units.

* *These goals are addressed in this section and are assessed in other sections of the unit.*

Pacing

- approximately four 45-minute class sessions

Vocabulary

- acre

About the Mathematics

The concept of finding area is first explored in the unit *Reallotment*, using methods such as counting squares and parts of squares, cutting and pasting parts of a shape to form a square or rectangle, and using formulas such as $A = length \times width$. The varied methods of finding one's own surface area introduced in this unit give students an opportunity to revisit area concepts at a more complex level.

Understanding the nature of square roots can be a challenging concept to grasp. It is important that students have experiences measuring areas with unusual square units, such as square furlongs, so that they develop a conceptual understanding of area units.

Another important concept that is introduced in this section is the use of a geometric model to model a real object. The surface area of the human body is compared to the surface area of two cylinders of equal height in the formula that students use on Student Book page 17.

Materials

- centimeter tape measures, pages 35, 39, 41 of the Teacher Guide (one per student or group of students)
- centimeter rulers, pages 35 and 39 of the Teacher Guide, (one per student)
- scissors, page 35 of the Teacher Guide, optional (one pair per student)
- tape, page 35 of the Teacher Guide, optional (one roll per group of students)
- newspapers, page 35 of the Teacher Guide (several per group of students)
- centimeter grid paper, page 39 of the Teacher Guide (one sheet per student)
- straightedges or rulers, page 43 of the Teacher Guide (one per student)

Planning Instruction

To introduce the context of this section, area, you might have a short class discussion about area. Ask students questions such as: *How does the area of our classroom compare to the area of your living room at home?* [Answers will vary. Many classrooms may be slightly larger than the average living room.] *Why might a person around your age be concerned with the area of the lawn surrounding his or her home?* [The child may be responsible for mowing the lawn surrounding his or her home.] *What units of measurement for area do you know?* [Answers will vary. Sample responses: square inches, square feet, square yards, square miles, square centimeters, square meters, square kilometers, and so on.]

You may want students to work on problems 2, 3, 7, 8, 13b, and 16 individually. Problems 9 and 10 may be done in small groups or individually and problems 1 and 13a are best done as whole-class activities. Students may work on the remaining problems in small groups.

There are no optional problems in this section.

Homework

Problems 2 (page 32 of the Teacher Guide), 7 and 8 (page 38 of the Teacher Guide), 10 (page 40 of the Teacher Guide), 13b (page 42 of the Teacher Guide), 16 (page 46 of the Teacher Guide), and the Extension activities (pages 35, 41, and 45 of the Teacher Guide) may be assigned as homework. After students complete Section B, you may assign appropriate activities from the Try This! section located on pages 35–37 in the *Made to Measure* Student Book as homework. The Try This! activities reinforce the main mathematical concepts covered in this section.

Planning Assessment

- Problem 16 can be used to informally assess students' ability to recognize that different units of measurement and different measurement systems exist; to recognize the importance of standard measurement units; to use formulas and rules of thumb to estimate and calculate area; to develop and use a frame of reference for measurement to solve problems; and to select and use appropriate measurement units.

B. AREAS

INTRODUCTION

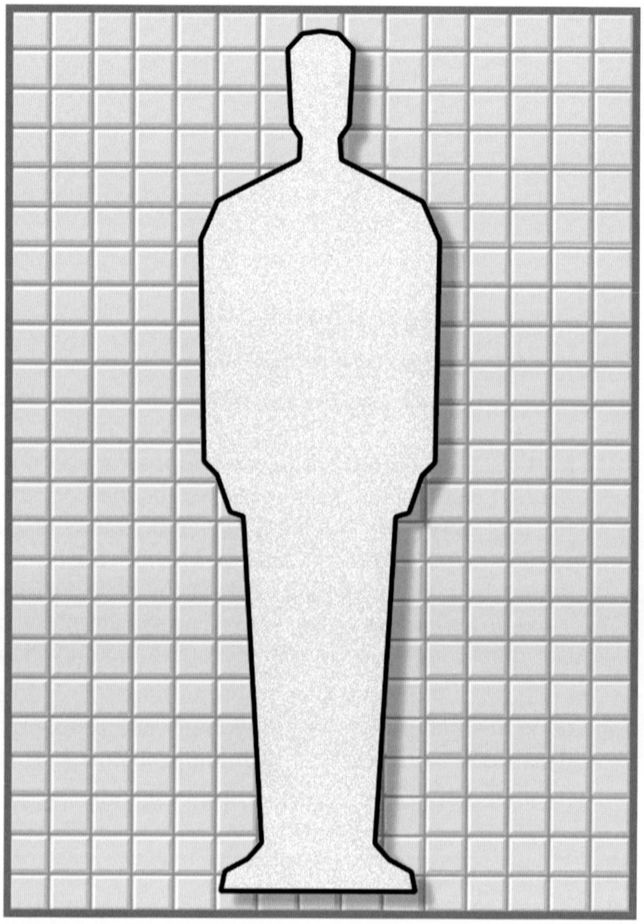

A body's *surface area* is the amount of skin that covers a body. Sometimes it is important to know the surface area of a person's body. For example, health care workers use estimates of a burned patient's body surface area to decide how much liquid is needed to replace the patient's lost fluids.

Body surface area is also important in understanding the needs of babies. A body cools down by sweating at a rate proportional to its skin area, but warms up in proportion to its body mass. A baby's skin area is big compared to his or her weight. Since babies cool down faster than adults, babies can feel uncomfortably cold even when adults feel warm.

1. How might you measure your body's surface area (or amount of skin)?

2. Estimate your body's surface area in square centimeters.

As you go through this section, you will estimate your body's surface area in many different ways. Make a table similar to the one shown below to keep track of your estimates.

Surface Area Estimation Table

Estimation Method	Estimation of Surface Area (in cm^2)

Every time you make a new estimate, record your answer in the table.

Solutions and Samples
of student work

1. Answers will vary. Sample responses:
   - I could use a rectangular sheet of paper to estimate the area of one side of my body and double the answer to include the other side.
   - I could wrap my body in newspaper and estimate the area of the sheets used.
   - I could estimate the number of sheets of paper of a particular size, such as self-stick notes or $8\frac{1}{2}$ by 11 sheets, that would cover my body and estimate the total area of the sheets used.

2. Answers will vary. Sample response:

   Surface area = 14,444 cm^2.
   Strategies will vary. Sample strategy using the rectangle method:

   Height: 5 ft 2 in. (157 centimeters) tall

   Shoulder to shoulder width: 18 inches (46 centimeters)

   Surface area = 157 × 46 × 2, or about 14,444 cm^2

Estimation Method	Estimation of Surface Area (in cm^2)
Student's own method: I made a rectangle to estimate my body's surface area and doubled the answer (to include my front and back).	157 cm × 46 cm × 2 = 14,444 cm^2

Hints and Comments

Overview In this section, students investigate various ways to estimate the surface area of their bodies. On this page, students think about ways in which they might measure their surface area and make a first guess of their surface area in square centimeters.

Planning Throughout this section, if students feel uncomfortable measuring themselves, you can let students work in small groups in which one student volunteers to be measured. Make sure, however, that every student does his or her own measuring. Students' personal measurements may be kept private. Measuring may be done using other units which should then be converted to square centimeters.

You might want to do problem **1** as a whole-class activity. Students may then work on problem **2** individually. Problem **2** may also be assigned as homework. Be sure to discuss students' answers for both problems.

Comments about the Problems

1–2. By estimating their body's surface area, students learn the meaning of surface area and the challenges of finding the surface areas of irregular shapes. The emphasis in these two problems is on students' thinking about possible methods with which to estimate their surface area.

2. Homework This problem may be assigned as homework. If students are uncomfortable lying on the floor, they may "wrap" themselves standing up instead.

For ease in comparing, all skin surface area measurements in this section should be recorded in square centimeters. Measuring may also be done using other units which are converted to square centimeters. *Note:* 5 centimeters is equal to 2 inches.

SQUARES

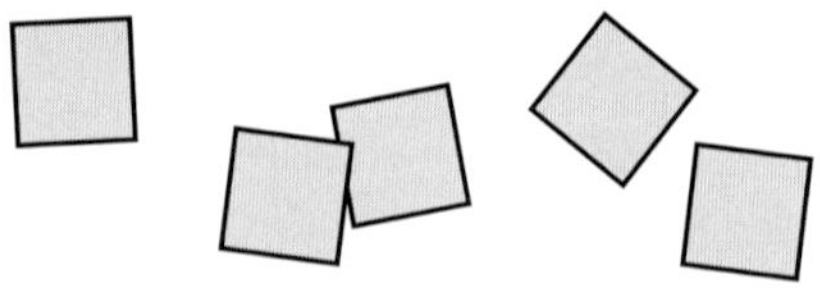

In Section A, you saw how you can fit in a square.

Of course, not everyone's square is the same size. You can use the square in which you fit to approximate your surface area. Three-fifths of your square's area is a good estimate of your surface area.

3. Explain why $\frac{3}{5}$ of the square is a good estimate of your body's surface area.

Activity

4. a. Make a life-size version of your square out of newspaper. (You'll probably need to tape some pages together.) Lie down on one end of the square and roll over one full rotation. Draw a straight line on the paper along the side of your body to mark the end of your roll. Did you roll over about $\frac{3}{5}$ of the square?

b. Shade $\frac{3}{5}$ of your newspaper square. Find the area of the shaded part in square centimeters. Record the result in your Surface Area Estimation Table.

c. To see how close your estimate is, you can try to "put on your skin." Using the newspaper square you made in problem **4a,** try to cover yourself. You may need to cut up your rectangle with scissors. Report how well your "skin" covers you.

Solutions and Samples
of student work

3. If you "roll" your body over the square, it rolls about $\frac{3}{5}$ of the way before you get back to where you started. The front and back of your body each take up about $\frac{1}{5}$ of the square, with your two sides together taking up another $\frac{1}{5}$.

4. a. Yes. See the solution for problem **3** for a sample explanation.

b. Answers and strategies will vary. Sample responses:

Strategy 1

I divided my square into five equal pieces and shaded three of the pieces. I divided 157 cm by five and multiplied by three to get the width (94.2 cm). Then I calculated the area of the shaded region:

$94.2 \times 157 = 14{,}789 \text{ cm}^2$

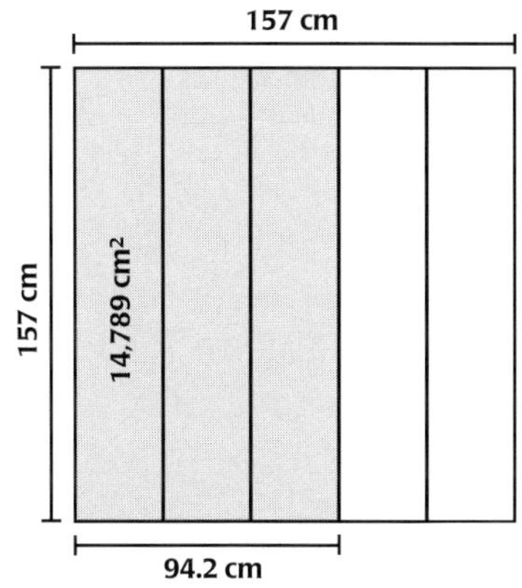

Strategy 2

I measured the length and width of the square and calculated its area:

$157 \times 157 = 24{,}649 \text{ cm}^2$

To find the area of the shaded region, I converted $\frac{3}{5}$ to a decimal (0.6) and calculated $\frac{3}{5}$ of $24{,}649 \text{ cm}^2$:

$0.6 \times 24{,}649 \text{ cm}^2 = 14{,}789 \text{ cm}^2$

Estimation Method	Estimation of Surface Area (in cm²)
I found the area of three-fifths of my personal square.	$\frac{3}{5} \times 157 \text{ cm} \times 157 \text{ cm} = 14{,}789 \text{ cm}^2$

c. Answers will vary, depending on how precisely students covered their body in the newspaper sections and their accuracy in calculating the individual areas and total area.

Hints and Comments

Materials newspapers (several per group); centimeter rulers or tape measures (one per group); scissors, optional (one pair per student); tape, optional (one roll per group)

Overview Students make life-size versions of their own squares. They then lie down and roll one full rotation to see if the area covered is about three-fifths of the square. Students shade and measure three-fifths of the square that approximates their body's surface area. Students then cut the shaded portion of the newspaper square into pieces and wrap them around their bodies to see how well this estimation method approximates their actual surface area.

Planning Students may work on problem **3** individually. Discuss students' answers to this problem before they work on problem **4** in small groups. Be sure that students record the results in their Surface Area Estimation Tables. Compare the results found by each group of students.

Comments about the Problems

4. If students are uncomfortable lying on the floor, they may "wrap" themselves standing up instead.

a. After students have approximated or calculated the area of their squares in square centimeters, they might use various methods to find three-fifths of the squares' area. Some students may divide the total area of the square by five and multiply that answer by three to find three-fifths of the area.

c. Remind students not to overlap the newspaper sections around their body parts. You might suggest that they use tape to secure the newspaper sections. Some students will cut and fit the newspaper sections in ways that fit better than methods used by other students.

Extension After students describe the methods they used to cover themselves with their "skin" in problem **4c,** ask them to suggest ways in which they could improve their methods to make a better fit.

There are other ways to find your body's surface area. Timm Ulrichs, a German artist, did many artistic experiments that were quite mathematical. Mr. Ulrichs found his own surface area in the following way:

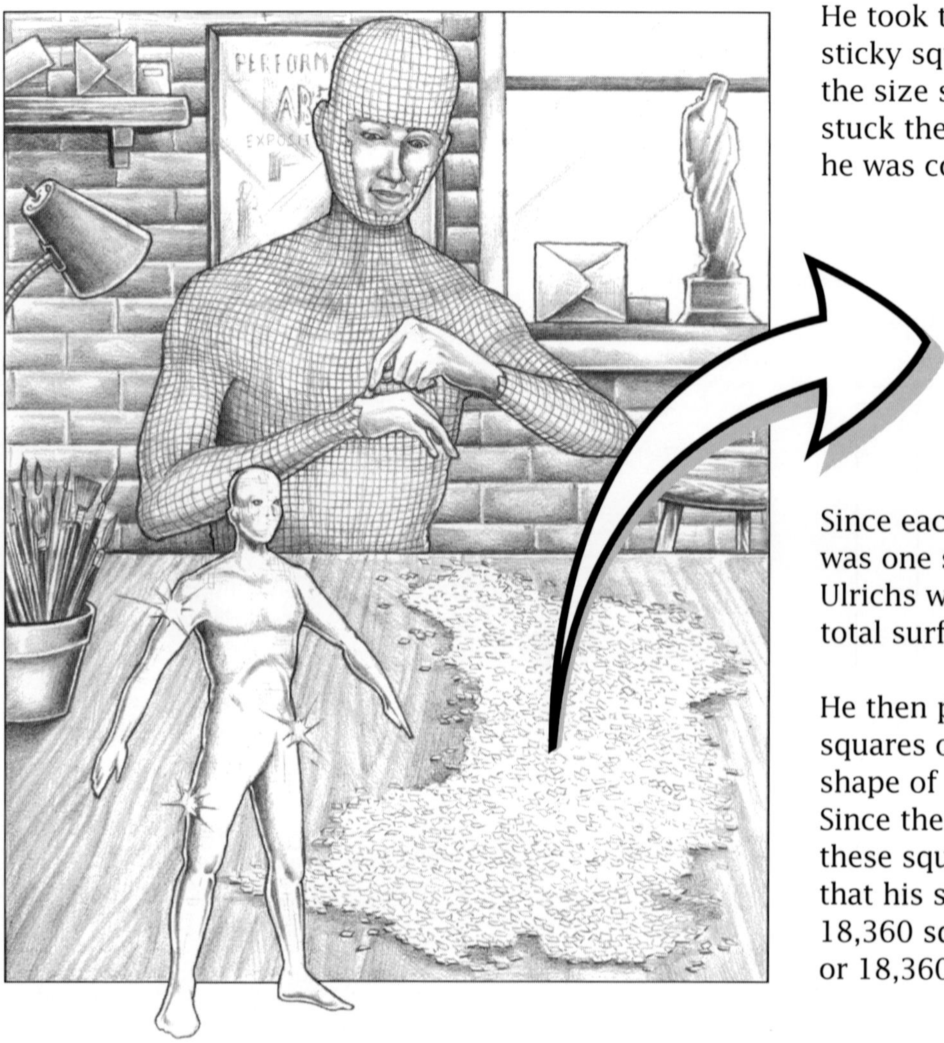

He took thousands of little sticky squares of foil (exactly the size shown below) and stuck them on his body, until he was completely covered.

1 cm²

Since each piece of foil was one square centimeter, Ulrichs was able to find the total surface area of his body.

He then put all of the foil squares on graph paper in the shape of a large rectangle. Since there were 18,360 of these squares, he concluded that his surface area was 18,360 square centimeters, or 18,360 cm^2.

5. Find some possible measurements for the length and width of Timm Ulrichs's rectangle. You don't have to be exact.

6. Find a rectangle with an area of about 18,360 cm^2, whose length could be Timm Ulrichs's height. Is this rectangle $\frac{3}{5}$ of Mr. Ulrichs's square? Explain your answer.

Solutions and Samples
of student work

5. Answers will vary. Sample responses that result in an area of about 18,000 cm^2:
   - 180 cm × 100 cm
   - 200 cm × 90 cm
   - 150 cm × 120 cm
   - 300 cm × 60 cm
   - 250 cm × 70 cm (area of 17,500 cm^2)

6. Answers will vary. Reasonable heights from the solution to problem **5** include: 150 cm, 180 cm, or 200 cm. Sample responses based on the above heights:

   If Mr. Ulrichs's height was 150 cm, then his square would measure 150 cm × 150 cm and $\frac{3}{5}$ of that is 13,500 cm (150 × 90) which is too small.

   If Mr. Ulrichs's height was 200 cm, then his square would measure 200 cm × 200 cm and $\frac{3}{5}$ of that is 24,000 cm (200 × 120) which is too large.

   If Mr. Ulrichs's height was 180 cm, then his square would measure 180 cm × 180 cm and $\frac{3}{5}$ of that is 19,440 cm (180 × 108) which is close to 18,000 cm, so this is a reasonable answer.

Hints and Comments

Overview Students read about an estimation method used by the German artist Timm Ulrichs to approximate his body's surface area. Students figure out possible dimensions of a rectangle that has the same area as Ulrichs.

About the Mathematics The relationships between length, width, and area are used in different ways:
area = *length* × *width* and *area* ÷ *width* = *length*.
Students use the second formula to determine a possible length, given the area of a rectangle.

Planning Students may work on problems **5** and **6** in small groups. Discuss students' solutions and strategies for both problems.

Comments about the Problems

5. You might suggest that students round the given surface area to 18,000 cm^2 to make it easier for them to find a possible length and width for the rectangle. To help students develop a frame of reference for this measurement, you might remind them that the dimensions 180 cm × 100 cm and 200 cm × 90 cm are close to 2 meters × 1 meter, the common dimensions of a door.

6. You might want to discuss the range of heights for average adult males in centimeters and then use a ratio table to make comparisons.

Inches	2	12	60	72
Centimeters	5	30	150	180

Students can either find approximate dimensions using their answers from problem 5 or they may divide the total area (18,000 cm2) by a chosen length, such as 200 cm, to find the width of the rectangle.

Hands and Body

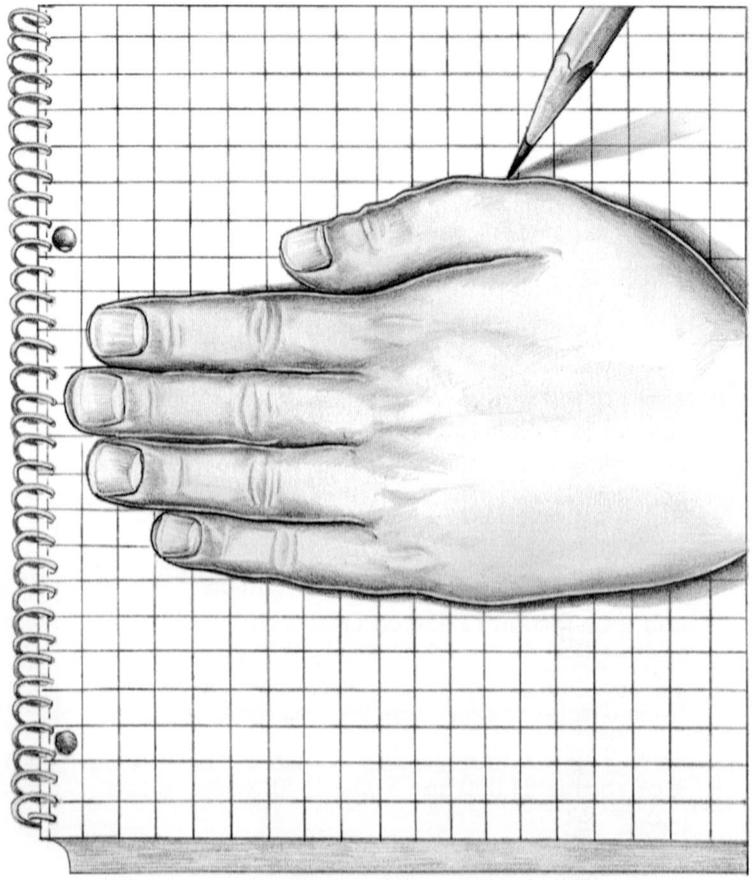

7. Trace your hand on grid paper as shown on the right. Estimate the area of your handprint in square centimeters.

Another method for estimating body surface area is to use the area of a handprint.

About 100 handprints cover the body.

8. Does the handprint method give the same estimate of your body's surface area as Timm Ulrichs's rectangle method? Record the result of the handprint method in your table.

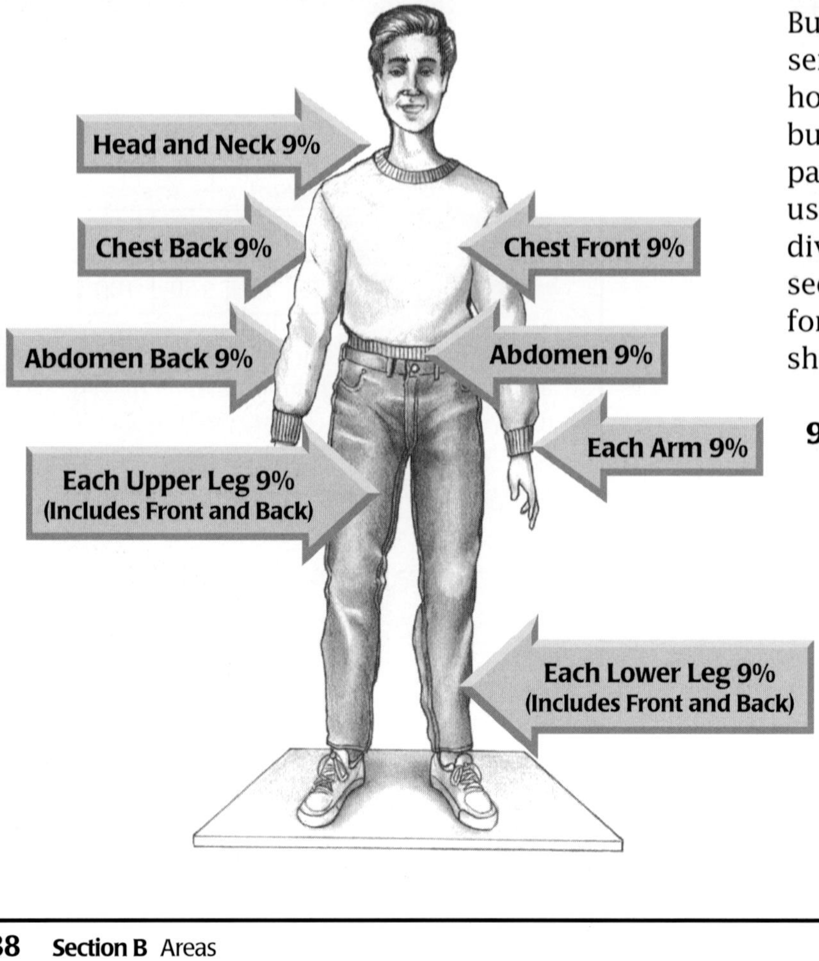

Burns can be very serious. The seriousness of a burn depends on how much of the body has been burned. To estimate the extent of a patient's burns, health care workers use the "rule of nines." This rule divides the body into eleven sections, each of which accounts for 9% of the total surface area, as shown in the picture on the left.

9. **a.** Think of a way to measure as precisely as possible one of the "rule of nines" sections of your own body surface. Measure the area of this section in square centimeters.

   b. Use the result to calculate your body's surface area. Record your body's surface area in your table from page 13 with your other estimates.

7. Answers will vary, depending on the size of students' hands. Sample response: 115 cm^2

8. No, the estimates are different. Sample response using the area of the handprint in the answer to problem **7:**

$100 \times 115 = 11{,}500$ cm^2. The handprint method resulted in a surface area estimate much smaller than the sample response using the rectangle method (14,789 cm^2).

See the solution for problem **9b** for sample estimation chart.

9. a. Answers will vary. Sample response:

The dimensions of my abdomen are about 28 cm by 41 cm. I multiplied to find its surface area: $28 \times 41 = 1{,}148$ cm^2.

b. Answers will vary. Sample response using the sample dimensions in part **a:**

Using the rule of nines, I multiplied the surface area of my abdomen times 11 to find my total body surface area: $11 \times 1{,}148 = 12{,}628$ cm^2.

Estimation Method	Estimation of Surface Area (in cm^2)
One hundred times the area of my handprint	100×115 cm^2 = 11,500 cm^2
Rules of Nines: I found the area of my abdomen and multiplied that answer by 11.	$11 \times (28$ cm $\times 41$ cm$)$ = 12,628 cm^2

Materials centimeter grid paper (one sheet per student); centimeter tape measures or rulers (one per student)

Overview Students trace their hand on centimeter grid paper and estimate the area of their handprint to determine their body surface area using the rule of thumb that their surface area is about the same area covered by that of 100 handprints.

Planning Problems **7** and **8** may be done individually as a homework assignment and discussed in class the next day. Make sure that students understand the "rule of nines" before they start taking their own measures in problem **9.** Students may work on problem **9** in small groups or individually.

Comments about the Problems

7–8. Homework These problems may be assigned as homework.

7. Be sure that students use centimeter grid paper to trace their hand to make it easier to estimate the area by simply counting the centimeter squares covered by their handprint.

8. You might ask students to compare the result of the "rectangle method" of estimating surface area used on Student Book page 14 with that of the "handprint method" used here. Ask: *Which method gave you a surface area that was smaller?* [the handprint method] *What might be some reasons that the two methods resulted in two different answers?* [The handprint method is a more precise measurement strategy.]

9. Students may use different methods with varying accuracy to measure each of the 11 sections of their body. Some sections are easier to find the area of than others. For example, it is easier to measure the surface area of an arm or leg than that of the head and neck. Some students may note that the percents add up to 99% rather than 100% ($11 \times 9\% = 99\%$). The difference of one percent can be ignored since the rule of nines is an estimation strategy.

Surface Area by Formula

There are both simple and complex formulas for finding the surface area of a person's body. One simple way is to multiply your height by your thigh circumference (the length around your thigh) and double the answer. The formula for this method is shown below.

$$\text{height (in cm)} \times \text{thigh circumference (in cm)} \times 2 = \text{body surface area (in cm}^2\text{)}$$

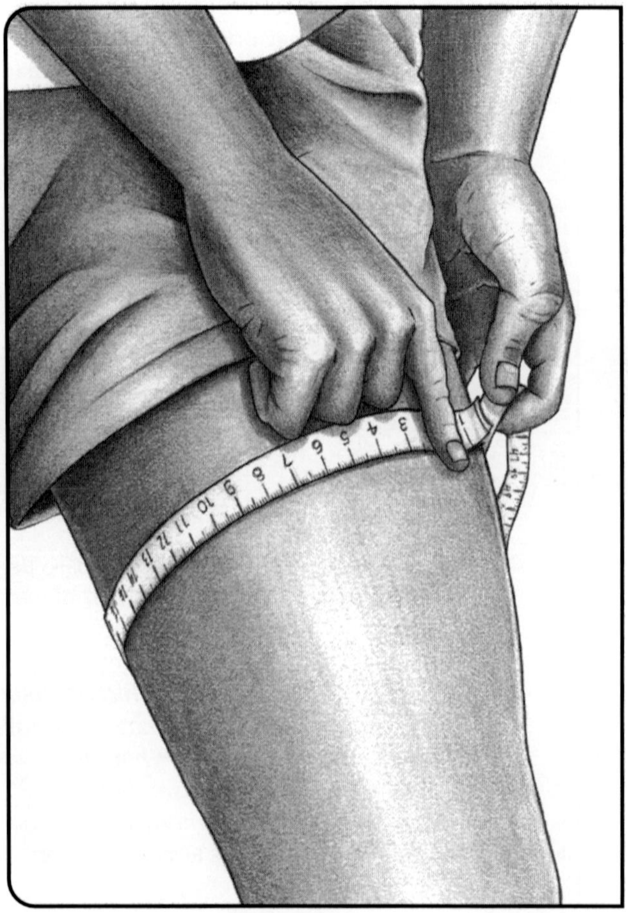

10. Measure your height and thigh circumference in centimeters, and then find your body's surface area using the above formula. Write the result in your Surface Area Estimation Table.

The above formula is based on the formula for the surface area of two cylinders.

11. Explain how the body can be seen as two cylinders of equal height.

12. Would the above formula work for babies? Explain why or why not.

10. Answers will vary. Sample response:

Height: 157 cm
Thigh circumference: 45 cm
Surface area: 157 × 45 × 2 = 14,130 cm^2

Estimation Method	Estimation of Surface Area (in cm^2)
I multiplied my height and the circumference of my thigh and multiplied the result by two.	157 cm × 45 cm × 2 = 14,130 cm^2

11. Since the height is used, the two cylinders are as tall as a person with a circumference of the size of one thigh. For a human body, the two cylinders encompass the legs and are as tall as the body. Since there are two legs, doubling the answer gives an estimate of the body's surface area.

12. No, the formula will probably not give a good estimation of a baby's surface area, since the thigh's circumference of a baby is much smaller than the torso and head.

Materials centimeter tape measures (one per student)

Overview Students measure their height and thigh circumference and find their surface area using a given formula. They explain how the body's surface area can be seen as equal to that of the two cylinders in the formula.

About the Mathematics Students first learn to calculate the surface area of cylinders in the unit *Reallotment.* An easy way to visualize or explain the surface area of a cylinder is to cut the cylinder and make it into a flat rectangle, the dimensions of which would be the person's height and the circumference of their thigh.

Planning Students may work on problem **10** individually or in small groups. Problem **10** may also be assigned as homework. Problems **11** and **12** may be completed in small groups. Discuss students' answers for all the problems.

Comments about the Problems

10. Homework This problem may be assigned as homework. If students are uncomfortable taking personal measurements, you might want them to take their measurements at home. As suggested earlier, students can also work in small groups in which one person volunteers to be measured.

11. Encourage students to think whether this is an accurate way to measure body surface area.

Extension Have students use the following formula to estimate the surface area of their bodies:

surface area = hip circumference × height

You might ask students to explain why this formula may give a more accurate approximation of the body's surface than the "two cylinder method" used above. [Since the hip circumference encompasses the outer circumference of the body, it results in a more accurate answer.]

Height, **Weight**, *and* Area

The surface area of a person's body is related to his or her height and weight. Unfortunately, there is no easy formula to find out exactly how these three measures are related.

Pictured on the right is a chart that health care workers often use to find the surface area of a patient's body. The chart is called a *nomogram.*

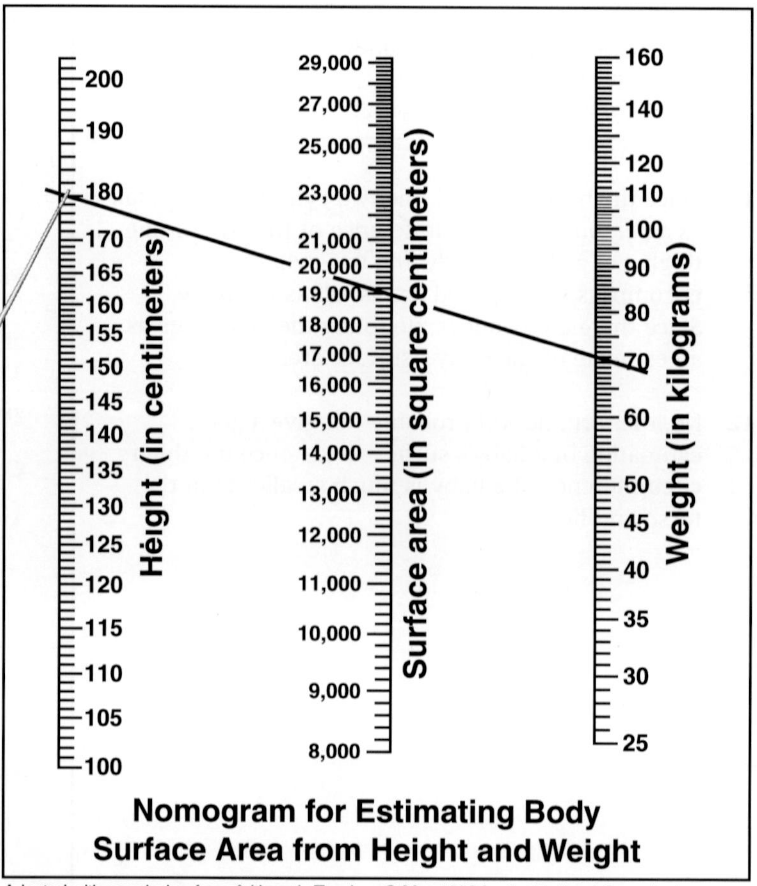

To use the above nomogram, first find the patient's height (in centimeters) in the left-hand column. Next, find his or her weight (in kilograms) in the right column. Then connect these two points with a line. The surface area of the patient's body is shown where the line crosses the middle scale.

13. a. A line for one patient has already been drawn on the nomogram. What information about the patient is indicated by the nomogram?

b. Use the nomogram and a straightedge to estimate your own body's surface area, and then enter the result in your table. (*Note:* 1 kilogram = 2.2 pounds)

13. a. The patient is 180 centimeters tall, weighs 70 kilograms, and has a body surface area of 19,000 cm^2.

b. Answers will vary, depending on students' heights and weights. Sample response:

A person weighing 61 kilograms who is 157 centimeters tall would have a body surface area of 16,200 cm^2.

Estimation Method	Estimation of Surface Area (in cm^2)
I used the nomogram chart to find my body surface area.	Weight: 61 kg Height: 157 cm Surface Area: 16,200 cm^2

Materials straightedges or rulers (one per student)

Overview Students use a nomogram chart and information about their heights and weights to determine their body surface areas.

Planning Students may work on problem **13a** as a whole-class activity. You may want students to work individually on problem **13b** and assign it as homework.

Comments about the Problems

13. b. Homework This problem may be assigned as homework. Have students measure their weights and heights at home. Students should measure their height in centimeters. Have students use a ratio table to change their weight in pounds to kilograms (22 lbs = 10 kg).

EARLY AREAS

In Section A, you saw that body measures were often used to find lengths in the past. At one time, area measures were also based on the human body. Square feet, square furlongs, and square thumbs were all used to measure area.

Area measures used in the United States were based on English measures. For areas, the English used such units as the *rood* and the *acre*. The acre is still used today.

In Section A, you discovered that a furlong is 660 feet long. Also, one square furlong equals 10 acres.

14. How long and how wide might 1 acre be? Express your answer in feet. Make a drawing to explain your answer.

A football field is about 100 yards long by 50 yards wide.

15. Use your answer to problem **14** to calculate how many acres are in a football field.

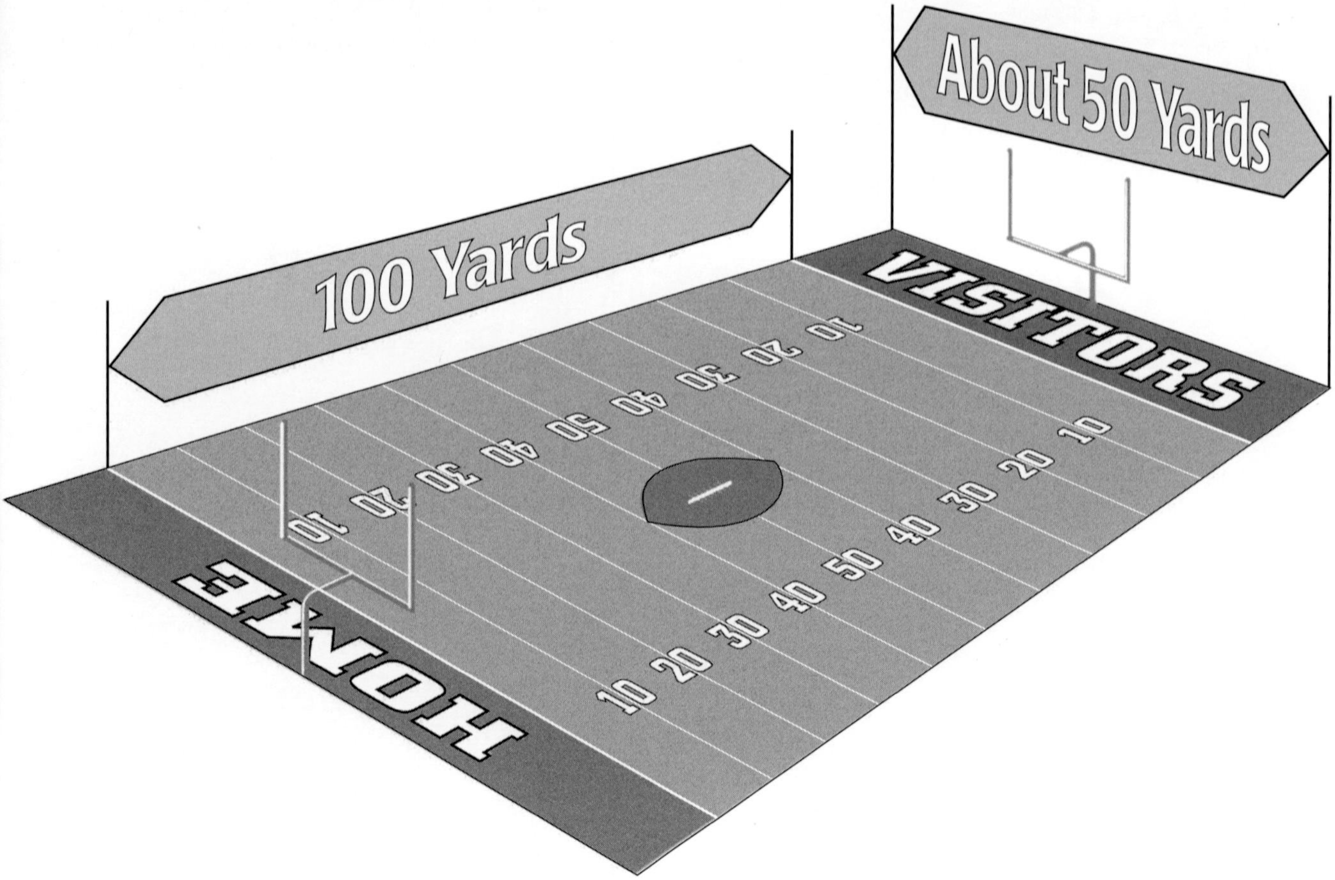

14. Answers will vary. Sample responses:

One acre is about 320 ft × 132 ft.

Students may first make a sketch of a square that represents 1 square furlong.

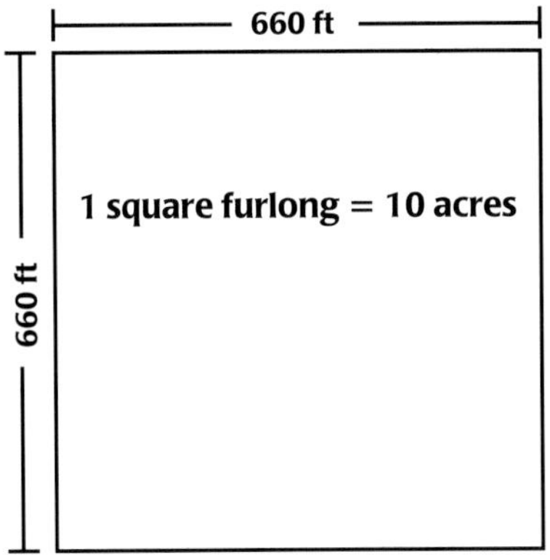

Secondly, students may divide the 660 ft × 660 ft square into 10 equal rectangles that each measure 330 ft × 132 ft.

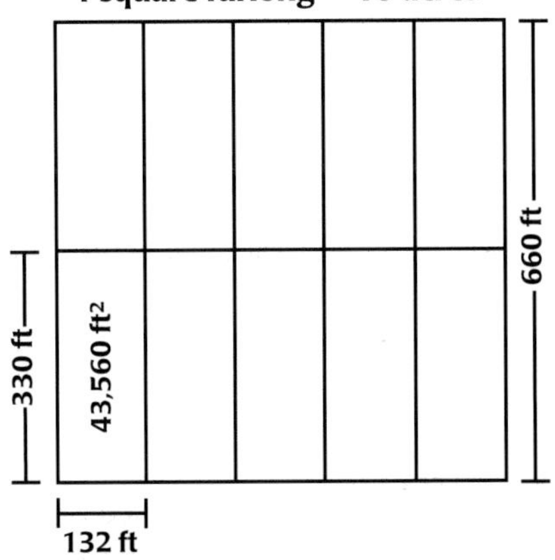

Thirdly, students might reason that a 660 ft × 660 ft square plot has a total of 435,600 square feet of area. If there are 10 acres in it, each acre will take up 43,560 square feet of land. They could then find some possible dimensions for the acre using a guess-and-check method. A square plot would be about 209 ft × 209 ft.

15. A football field is about 1 acre. Sample student response:

A football field measures 300 ft by 150 ft, so it has a total area of 45,000 ft^2 (300 × 150 = 45,000 ft^2). Since an acre has an area of 43,560 ft^2, a football field takes up about the same space as 1 acre.

Hints and Comments

Overview Students learn about historic area measurement units based on the human body, such as square furlongs and square thumbs. They explore the area unit acre (a square unit) and determine how many acres are in a football field.

Planning Students may work on problems **14** and **15** in small groups. Be sure to discuss students' answers for both problems.

Comments about the Problems

14–15. The purpose of these problems is to help students get a sense of the size of an acre. They relate the unit acre to the dimensions and area of a football field in order to develop a frame of reference for measurement.

Extension Have students measure the dimensions and calculate the area of the plot of land on which their house or apartment is built. Many such land plots are one-fourth acre in area. Have students try to imagine the areas of their property compared to that of a football field. Students might also compare the areas of the school parking lot or of a soccer field to the area of a football field and acre.

Summary

Surface areas are important for figuring out the amount of material needed to cover something. Body surface areas are used to determine the amount of fluid needed by burn victims.

In this section, you learned about several different methods to find your body's surface area. These methods used squares ($\frac{3}{5}$ of your square), handprints (about 100 handprints), a formula based on the area of two cylinders (height × thigh circumference × 2), and a nomogram.

Summary Questions

16. Write a paragraph or two comparing the different methods you used for estimating body surface area. Which of these methods do you think is most accurate? Why?

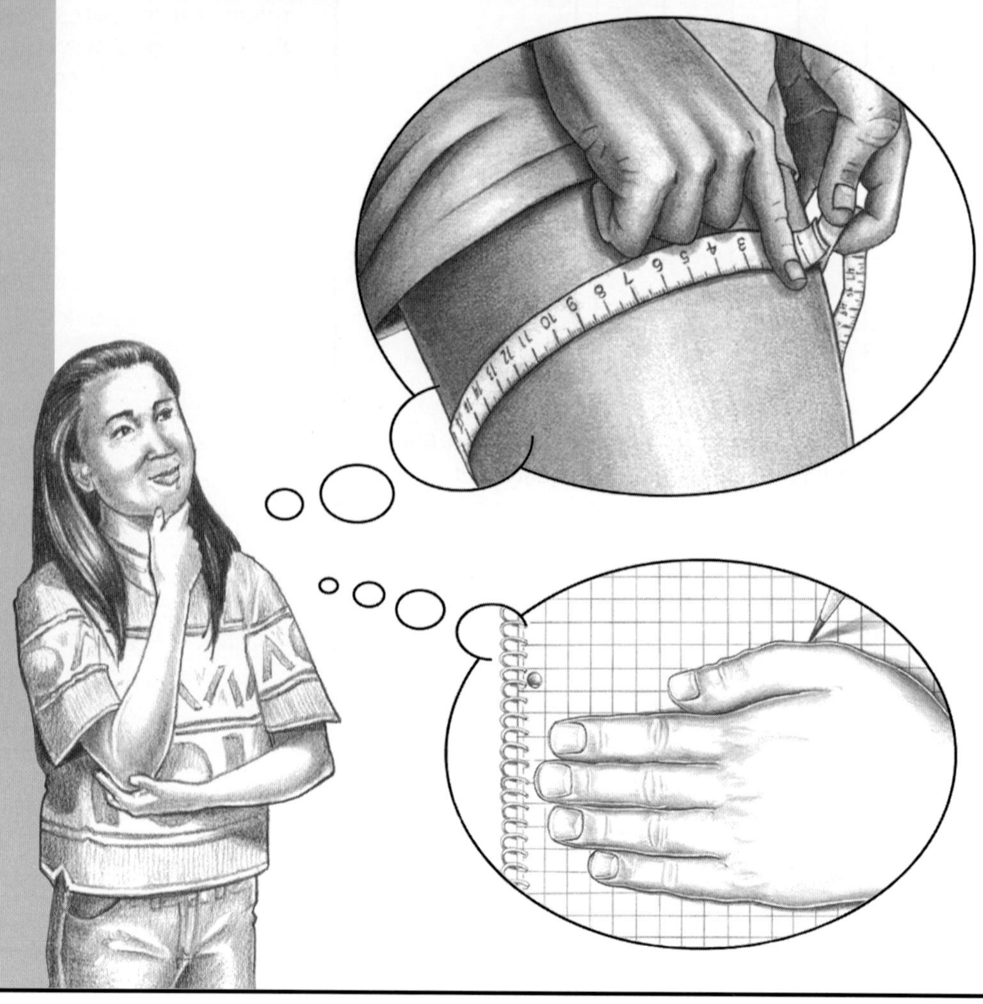

Solutions and Samples
of student work

16. Answers will vary. Students' paragraphs may include a description of each method. They might also indicate which was their favorite method, or the method that they felt was the most accurate. They might also find the average of the six estimation methods discussed in this section.

Hints and Comments

Overview Students read and discuss the Summary, which reviews the main estimation methods used to find a person's body surface area. Students write a paragraph that compares the estimation methods and comment on which method they think is the most accurate.

Planning Read and discuss the Summary with students. You might want them to look through this section once more to review the different area estimation methods. Problem **16** may be used as an informal assessment and/or assigned as homework. Students may use their Surface Area Estimation Tables to help them compare the various methods. You may want students to work on this problem individually.

Comments about the Problems

16. Informal Assessment This problem assesses students' ability to recognize that different units of measurement and different measurement systems exist; to recognize the importance of standard measurement units; and to use formulas and rules of thumb to calculate and estimate area. It also assesses their ability to develop and use a frame of reference for measurement to solve problems and their ability to select and use appropriate measurement units. This problem may also be assigned as homework.

Note: The nomogram probably gives the most accurate information since this reference chart is used in medicine.

Work Students Do

Students examine liquid and solid volume measurements of a variety of objects, including some related to the human body. First, students investigate solid volume measurements by examining boxes. They learn that the volume of a box may be found by multiplying the base area by the height. Next, students examine liquid measurements. Then they find a relationship between liquid and solid measurements. Students estimate their own body volume by using cubes to model the body, and they learn about some historical volume measures that are related to the human body.

Goals

Students will:

- estimate lengths, areas, volumes, and angles using appropriate units;
- find (compute, measure, and draw) lengths, areas, volumes, and angles using standard and nonstandard units;
- use simple relationships between and within measurement systems;
- recognize that different units of measurement and different measurement systems exist;
- recognize the importance of standard (measurement) units;*
- understand and use the relationships between length, area, volume, and angle;
- use geometric models to solve problems;*
- use formulas and rules of thumb to estimate and calculate length, area, and volume;
- develop and use a frame of reference for measurement to solve problems;*
- select and use appropriate measurement units.

** These goals are assessed in others sections of the unit.*

Pacing

- approximately four 45-minute class sessions

Vocabulary

- cup, pint, quart, gallon
- milliliter, liter
- cubic centimeter, cubic decimeter, cubic meter
- cubic cubit, cubic fathom,
- full cord and face cord

About the Mathematics

Volume measurements can be done in both liquid measures (for instance, milliliters, deciliters, and liters; pints, quarts, and gallons) and solid measures (for instance, cubic centimeters and cubic decimeters; cubic inches and cubic feet). In the metric system, the conversions between the two types of volume units are easy: 1 cubic centimeter = 1 milliliter, 1 cubic decimeter = 1 liter. In customary units, used in the United States, the relationships between units are not easy: for example, 1 quart = 57.73 cubic inches.

Solid volumes can be found by counting cubes or by multiplying the base area by the height. The formula *base area times height* is perhaps better than the formula *length times width times height* because of its greater generality. For example, a cylinder has a volume that is the area of the circular base (πr^2) times the height.

Encourage students to distinguish between *area* and *volume* measures. Area is two-dimensional and is needed to cover an object. Volume is three-dimensional and is needed to fill an object.

Materials

- centimeter grid paper, page 53 of the Teacher Guide (one sheet per student)
- cubic decimeter made from a net, page 53 of the Teacher Guide, optional (one per class)
- tissue boxes, page 53 of the Teacher Guide (one per group of students)
- centimeter rulers, page 55 of the Teacher Guide (one per student)
- $\frac{1}{2}$-liter, 1-liter, and 2-liter bottles, page 57 of the Teacher Guide, optional (one of each per class)
- orange juice carton, ketchup bottle, mayonnaise jar, and small carton of milk, page 57 of the Teacher Guide (one of each per group of students)
- metric measuring cups or beakers, pages 57 and 59 of the Teacher Guide (one per class or per group of students)
- centimeter tape measures, page 61 of the Teacher Guide (one per group of students)
- metric tape measures and beakers of various sizes, page 65 of the Teacher Guide (several per class)

Planning Instruction

To determine students' initial understanding about solid and liquid measurements, you can ask students to estimate the volumes of the cans and bottles that soda comes in and to compare sizes. Make rough drawings on the board or bring in the cans and bottles. Ask for estimates in liters or milliliters and in cubic centimeters.

Students may work on problems 1, 2b-11, 13, and 15-19 in small groups. Students may work on problems 2a, 12, 20, and 21 individually. Problem 14 may be done as a class.

There are no optional problems in this section.

Homework

Problem 7 (page 54 of the Teacher Guide) and the Bringing Math Home activity (page 57 of the Teacher Guide) can be assigned as homework. After students complete Section C, you may assign appropriate activities from the Try This! section, located on pages 35–37 in the *Made to Measure* Student Book, as homework. The Try This! activities reinforce the main mathematical concepts covered in this section.

Planning Assessment

- Problem 20 can be used to informally assess students' ability to estimate lengths, areas, and volumes using appropriate units; to find (compute, measure, and draw) lengths, areas, volumes, and angles using standard and nonstandard units; and to recognize that different units of measurement and different measurement systems exist. It also assesses their understanding of and ability to use the relationships between length, area, and volume; as well as their ability to use formulas and rules of thumb to estimate and calculate length, area, and volume and to select and use appropriate measurement units.
- Problem 21 can be used to informally assess students' ability to use simple relationships between and within measurement systems.

C. VOLUMES

INTRODUCTION

In the last two sections, you measured the lengths and surface areas of different parts of your body. You can also measure, estimate, and calculate volumes. Knowing the volume of a three-dimensional object is useful when you want to describe how much space it takes up.

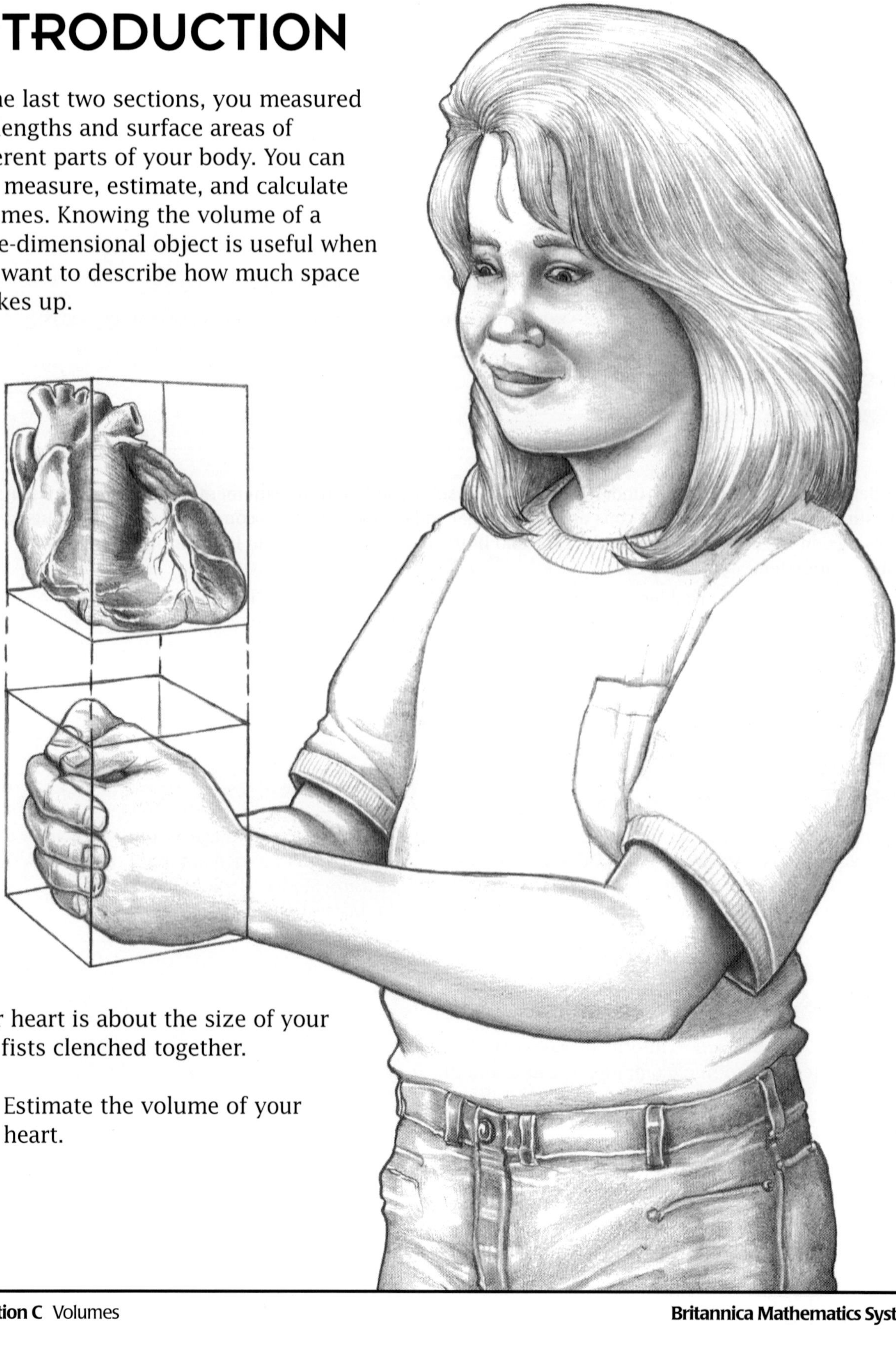

Your heart is about the size of your two fists clenched together.

1. Estimate the volume of your heart.

Solutions and Samples
of student work

1. Answers will vary. Sample student responses:
   - 1–2 cups
   - a pint
   - about the volume of a grapefruit

Hints and Comments

Overview Students read the introduction to this section. They learn that their heart is about the size of their two fists clenched together, and they are asked to estimate its volume.

Planning Students may work on problem **1** in small groups. Discuss their answers in class.

Comments about the Problems

1. This problem is critical because it is the first time that students are asked to think about volume. This would be a good place to discuss the different ways that students measure volume. Some students may use liquid measurements, some may use solid measurements, and some may use analogies to describe the size of the heart.

   Students may come up with nonstandard units of measure or standard units of measure, such as cubic centimeters, milliliters, cubic inches, and so on.

   You might ask students whether they know of any relationships between the different measures they come up with. It is not important for students to "look up" relationships, however. It is more important for students to get a general sense of the sizes of various units of volume, both liquid and solid. This is a major focus of this section.

SOLIDS

2. a. You can make a centimeter cube (1 cm^3) by drawing a figure like the one on the right and then folding it up into a cube. The figure is called the *net* of a cube. Use centimeter grid paper to draw a net of a centimeter cube. Cut it out and then fold and tape it to make one cubic centimeter.

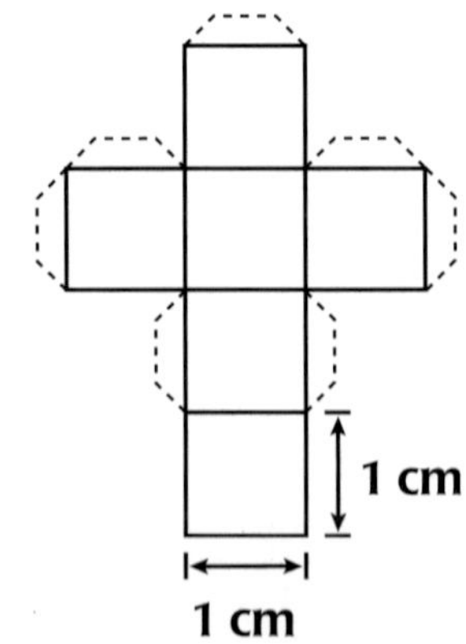

   b. Are all of the centimeter cubes made by your class enough to fill up one cubic decimeter? (*Note:* there are 10 centimeters in one decimeter.)

   c. How many centimeter cubes are needed to completely fill one cubic decimeter?

3. Cut the top off a tissue box. How many cubic centimeters are needed to fill it? Explain how you found your answer.

If it takes 32 centimeter cubes to completely fill a box, we say that the volume of the box is 32 cubic centimeters. The volume of larger objects can be expressed in cubic decimeters or cubic meters (in the metric system) or in cubic feet or cubic yards (in the English system).

4. a. For what kinds of objects would volume be measured in cubic centimeters? cubic decimeters? cubic meters?

   b. Write some statements about how cubic centimeters, cubic decimeters, and cubic meters are related.

Maria is trying to find the volume of the box shown below. The box has been partially filled with cubes.

Maria says, "I can easily find the volume of the box. Since the box measures 8 centimeters by 7 centimeters, I can fit 56 cubes on the bottom layer."

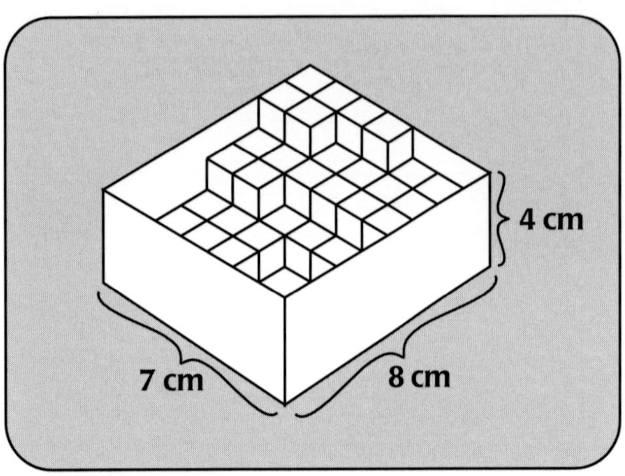

5. Explain what else Maria has to do to find the volume of the box.

6. Jonathan has a fragile vase that he is mailing in a box. To cushion the vase, he bought a box of packing material that says "Will fill a box as big as 6,000 cubic centimeters!" His box is 35 centimeters by 16 centimeters and has a height of 10 centimeters. Will there be enough packing material to fill the box? Show your work.

Solutions and Samples

of student work

2. a. Check to see that students correctly made a model of a cubic centimeter.

b. No. It would take 1,000 cubic centimeters to fill a cubic decimeter (which is 10 cm × 10 cm × 10 cm). Cubes made by one class would not even fill a single layer.

c. It takes 1,000 cubic centimeters to fill a cubic decimeter.

3. Answers will vary. Sample student response:

My tissue box measures 5 cm × 11 cm × 23 cm. The bottom layer requires 253 cubes and with five layers, there would be 253 × 5 or 1,265 cubes.

4. a. Answers will vary. Sample responses:

- cubic centimeters: small boxes, bags of flour or sugar
- cubic decimeters: large boxes, cans, milk cartons
- cubic meters: houses, grain, coal

b. Statements may vary. Sample response:

1 dm^3 = 1,000 cm^3. Since there are 10 centimeters in a decimeter, there are 10 × 10 × 10 = 1,000 cm^3 in a dm^3.

1 m^3 = 1,000 dm^3. Since there are 10 decimeters in a meter, there are 10 × 10 × 10 = 1,000 dm^3 in 1 m^3.

1 m^3 = 1,000,000 cm^3. Since there are 1,000 cm^3 in a cubic decimeter, there are 1,000,000 cm^3 in a cubic meter.

5. Maria has to multiply the 56 cubes in the bottom layer by four, which is the total number of layers (4 × 56 = 224 cm^3).

6. Yes, more than enough. The volume of one layer of the box is 560 cm^3 (35 × 16). So, since the box can hold 10 layers, its volume is 5,600 cm^2. The packing material will fill 6,000 cm^3, so the packing material will fill the box (especially since the vase must be packed as well.

Hints and Comments

Materials centimeter grid paper (one sheet per student); cubic decimeter made from a net, optional (one per class); tissue boxes (one per group of students)

Overview Students make a cubic centimeter from a net and investigate how many are needed to fill a cubic decimeter. They think about the appropriateness of measuring objects in different metric units.

About the Mathematics For rectangular boxes, the volume may be calculated using the formula: *volume* = *length* × *width* × *height*. Another formula that may be used is *volume* = *base area* × *height*. The latter formula may also be used for other objects, like cylinders and prisms. Note that it cannot be used for pyramids or conical objects.

In the metric system, volume is measured in cubic centimeters, cubic decimeters, and cubic meters. A cubic decimeter is the same as one liter, which is used for liquids.

Planning Students should work on problem **2a** individually. They may work in small groups on problems **2b–6.** Discuss students' answers for all the problems.

Comments about the Problems

2. b. You may wish to make a cubic decimeter so that students can place their cubic centimeters in it to get a better grasp of the volume in question.

3. Students should realize that they do not need to fill the box with cubic centimeters. They may measure the dimensions of the box in centimeters and use a formula for volume. Or they can use the cubic centimeters they made to find out how many fit along the length and the width of the box (multiplying these numbers will give the number of cubic centimeters of one layer) and then again using a cubic centimeter to find out how many layers will fit into the box.

4. a. Students do not have to calculate the volumes of the objects they name. They are merely asked to describe which objects are appropriately measured in the given units.

7. Margaret has four different boxes. Each one has a volume of 18 cubic decimeters. What might the measurements of these boxes be?

8. Find the volume of each item below and describe your strategy.

a.

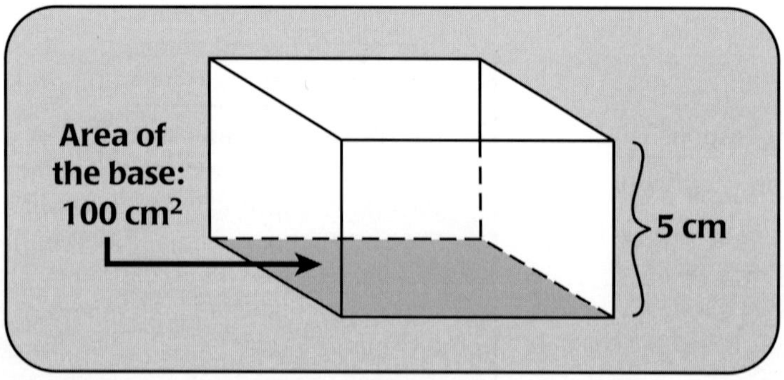

b. a deck of cards

c. a soda can (*Hint:* What is the area of the bottom of the can?)

9. Suppose that you want to determine the volume of your classroom. Can you use the area of one wall to find the volume of your classroom? Find the volume of your classroom.

10. Is it possible to use centimeter cubes (like the one you made on page 22) to find the volume of a soda can? Explain your reasoning.

You can often find the volume of an object by using the following formula:

volume = *area of the base* × *height*

11. The above formula does not always give the correct answer for all three-dimensional objects. For example, the volume of the pyramid shown on the right cannot be found by using this formula.

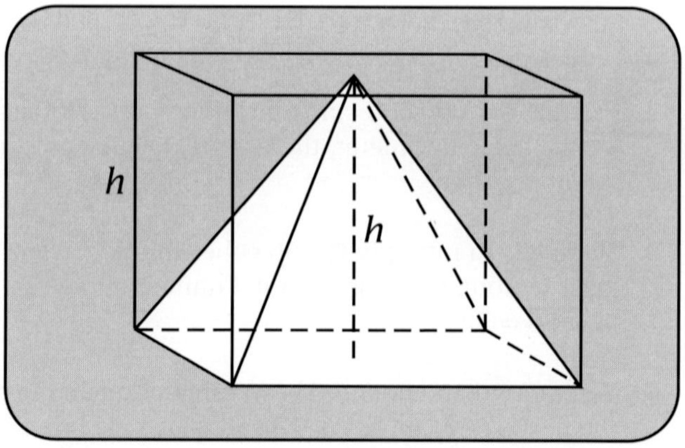

a. Give an example for which the above formula does work and another example for which the formula does not work.

b. Why does the formula give the correct answer for some objects but not for others?

You can use the area of your handprint (Section B, problem **7**) to estimate the volume of your hand.

12. a. Besides the area of your handprint, what other measure(s) do you need to estimate the volume of your hand?

b. Estimate your hand's volume.

7. Answers will vary. Sample response:

- 1 dm × 1 dm × 18 dm
- 2 dm × 3 dm × 3 dm
- 2 dm × 2 dm × 4.5 dm
- 2 dm × 9 dm × 1 dm

8. a. 500 cm^3. Strategies will vary. Sample student strategy:

The area of the base is 100cm^2. This means that there are 100 cubes that make up the bottom layer. The box will hold 5 layers of 100 cubes, so its volume is 500 cubic centimeters.

b. 108 cm^3. The dimensions of a deck of cards is 6 cm by 9 cm by 2 cm, so its volume is 108 cm^3 (6 × 9 × 2).

c. Answers will vary depending on the size of the can. Sample response:

By counting squares on grid paper, I found that the area of the circle is about 35 cm^2. Since the can is 12 cm high, the total volume is about 420 cm^3.

9. Yes, the area of the wall times the length of the floor will give the volume. Students may use meters or feet in measuring, so the volume will be in cubic meters or cubic feet.

10. Answers will vary. Some students might say no, noting that cubes cannot fill the entire volume since the can is not a block. Other students might say yes, if the cubes are cut into pieces and fit into the can. Or they may suggest ways of estimating with cubes.

11. a–b. Examples will vary. The area of the base tells how many cubes are needed for the bottom layer, and the height tells how many layers you need. This works only if the areas for all "layers" in the object are the same. If the object is a pyramid or a cone, or a conical or irregular shape, this formula will not work.

12. a. The thickness of the hand.

b. Estimates will vary. Students may note that the thickness of the hand varies from the fingertips to the wrist. Using the estimate of the surface area of the hand from Section B (115 cm^2), the volume of the hand is about 130 cm^3.

Materials centimeter rulers (one per student)

Overview Students investigate the formula *volume = area of the base × height* and explain why it works for some objects but not for others.

Planning Students may work on problems **7–11** in small groups. Discuss their answers in class, especially those to problem **11.** Students may work on problem **12** individually. Problem **7** may be assigned as homework.

Comments about the Problems

7. Homework This problem may be assigned as homework. Ask students to describe the dimensions of the boxes by comparing them to the dimensions of objects they know, like a desk, a hand span, and so on.

8. b. You might ask students whether the volume changes when the deck of cards is "slanted."

c. Students will probably need to trace the base on grid paper and count the squares. Students are not expected to use the formula $A = \pi r^2$. Note that the volume of the liquid is 355 milliliters (a milliliter is equal to a cubic centimeter), which is smaller than the volume for the can.

10. The intent of this question is to help students realize the limitations of using "strict" cubic centimeters as measuring tools and to see that sometimes cubic centimeters need to be divided (see also problem **6**).

11. This problem formalizes the volume formula (*volume = base area × height*) and the concept of using consistent measuring units. Be sure that students understand when the given volume formula is applicable. Have students draw or cut out pictures of shapes for which the formula will work and shapes for which the formula will not work.

12. b. To determine the depth of their hand, students may decide to measure the thickness in several established places and then take an average of those measurements.

LIQUIDS

The units used to measure liquids are different from the units used to measure solids. Pints, quarts, gallons, liters, and milliliters are all typical measures of liquid volume.

Shown on the right are some bottles whose contents are measured with units of liquid volume.

You might find it helpful to know that

- a regular soda can contains about $\frac{1}{3}$ liter.
- a cubic decimeter contains exactly 1 liter of liquid.

TRY THESE!

13. a. Estimate the volume (in liters) of the objects shown on the right.

b. Estimate the volume (in liters) of the following objects:

- a small container of juice
- a large cup of soda
- a small cup of coffee
- a gallon of milk

Solutions and Samples
of student work

13. a. Answers will vary. Sample response:

- orange juice carton: about 2 liters
- ketchup bottle: about $\frac{1}{2}$ liter
- mayonnaise jar: about 1 liter
- small milk carton: about $\frac{1}{4}$ liter

b. Answers will vary. Sample response:

- small container of juice: about $\frac{1}{3}$ liter
- large cup of soda: about 1 liter
- small cup of coffee: $\frac{1}{4}$ liter
- one gallon of milk: about 4 liters

Hints and Comments

Materials $\frac{1}{2}$-liter, 1-liter, and 2-liter bottles, optional (one of each per class); orange juice carton, ketchup bottle, mayonnaise jar, and small carton of milk (one of each per group of students); a metric measuring cup or beaker, optional (one per class)

Overview Students learn that measurement units for liquids are different from units used to measure solids. They estimate the volumes, in liters, of food packages.

Planning Have students read the introductory text. Then discuss with the class how liquids are measured differently than solids. Students may work on problem **13** in small groups. Discuss students' solution strategies in class.

Comments about the Problems

13.a–b. Students need the actual objects that appear on Student Book page 24 to find the volume of each item. They may have to convert pints and gallons to liters, or they may estimate the volume in liters.

b. Students may use either the actual objects or their findings from problem **13a** as reference measures.

Extension You may wish to bring to class a beaker that shows milliliters to measure the volumes of containers. Fill some empty jars with water, have students estimate the volume of each jar in liters, and then pour the contents into the beaker to determine the actual volume.

Bringing Math Home Students may investigate the volumes (in liters) of containers that they find at home and then write reports on their findings.

Activity

Measuring the Volume of Your Hand

For this activity, you will need a can or beaker that measures in milliliters. It should be big enough so that you can put your hand in it.

Pour water into the can or beaker so that it is about half full. Measure and record the level of the water. Then, put a rubber band on your wrist, and put your entire hand in the water up to the bottom edge of the rubber band. You may want to make a fist. Be sure that your whole hand is underwater.

Measure and record the new water level. The difference between the old level and the new level is the volume of your hand in milliliters.

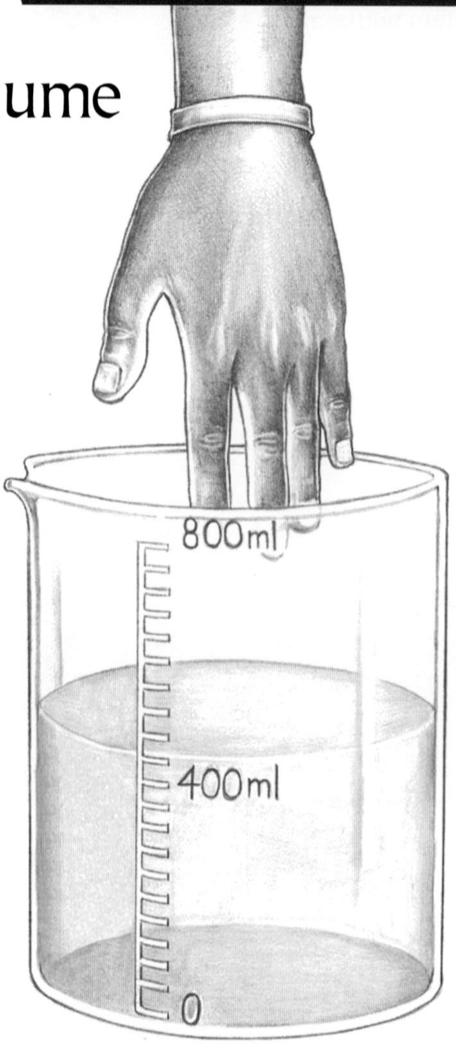

You now have two estimates of the volume of your hand: one in cubic centimeters (from problem **12**) and the other in milliliters. How are these two measurements related?

14. Make a table listing the two estimates of hand volumes for 10 students. Find a relationship between the two measurements. Drawing a graph might be helpful.

14. Answers will vary. Sample table:

Name of Student	Volume of Hand in cm3	Volume of the Hand in ml

The volume in cubic centimeters is the same as the volume in milliliters since 1 milliliter of water occupies 1 cubic centimeter of space. The points on the graph should show a strong linear relationship (in fact, the graph should be roughly a line with slope 1 and y-intercept 0). If differences exist, they may be due to errors in measurement. (The volume will probably be about 200 cm^3.) Sample graph:

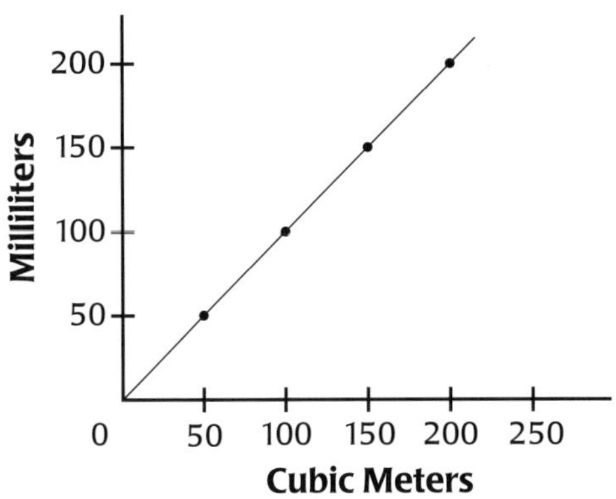

Materials metric measuring cups or beakers (one per group)

Overview Students measure the volume of their hand by immersing their hand in water in a measuring cup or beaker. They investigate the relationship between hand volume in cubic centimeters and hand volume in milliliters.

Planning Students may do this activity in small groups. Make the table in problem **14** as a whole-class activity, and have students work in small groups to discover the relationship between the two measurements.

Comments about the Problems

14. By this point in the unit, students should realize that 1 milliliter = 1 cubic centimeter, and that there is a relationship between liquid size and solid size, although they are measured in units that have different names. You may want to remind students that one cubic decimeter is the same as one liter.

Ask the science teacher at your school for beakers that are large enough for students to immerse their hands into (1,000-ml beakers will work). If large beakers are unavailable, 2-pound coffee cans will work. Various water levels can be determined and marked on the inside of the can by using a graduated cylinder and waterproof marking pen.

Ask students whether they think the answer would change if they made their hand into a fist. Some students may need help creating the axes for the graph.

Extension Weight is not a topic in this unit, but you may also want to mention to students that 1 liter of water (which equals 1 cubic decimeter) weighs one kilogram (and therefore 1 milliliter of water weighs 1 gram).

The **Volume** of Your Body

You could find the volume of your body in the same way that you found the volume of your hand, using liquid measures, but you would probably need a bathtub to do it!

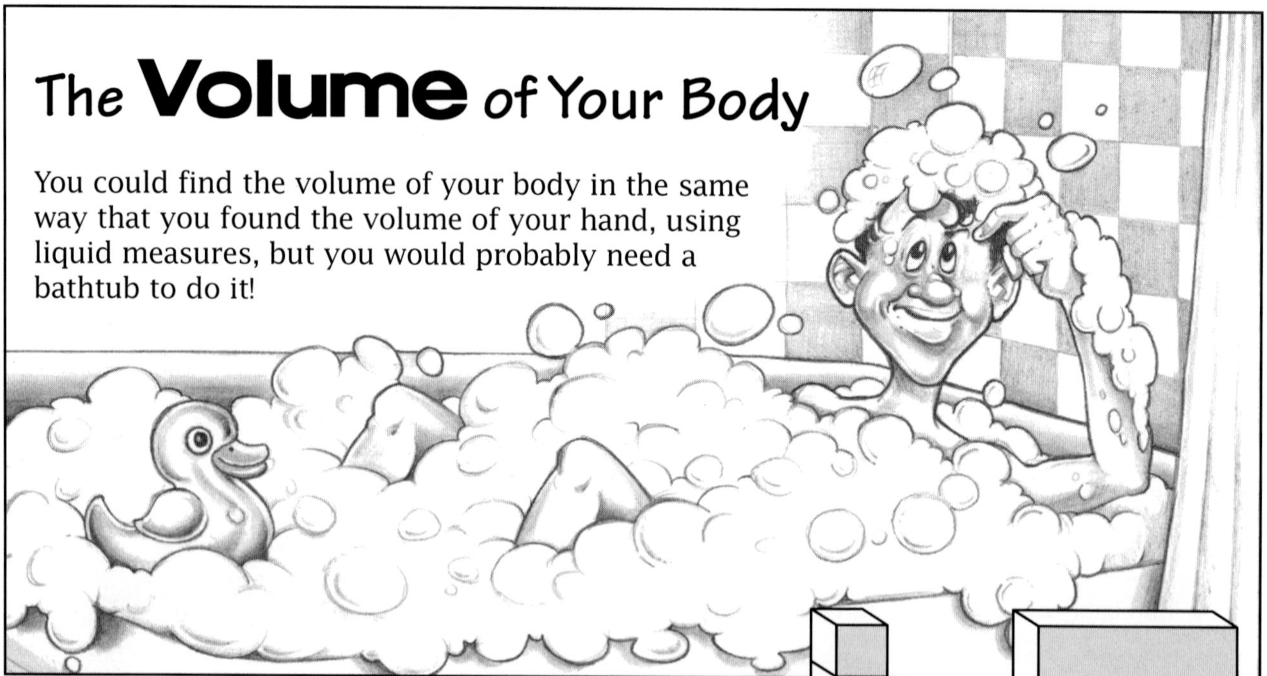

Another way is to use solid units of measure and to think of your body as being made up of cubes

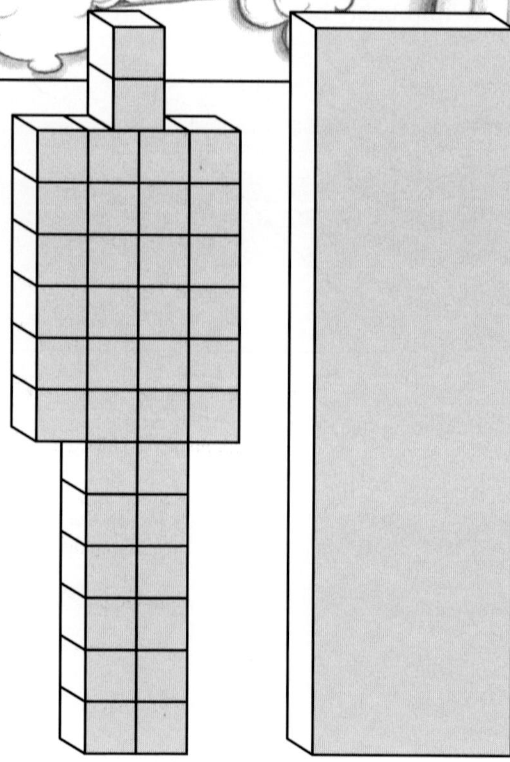

15. Estimate your body's volume by modeling your body with cubes as shown on the right. You will need to decide on a good unit of measure for the cubes.

To estimate the volume of your body, you can also use a block to model your body, as shown on the far right.

The table below contains some information about a girl at ages 10, 13, and 16.

Data for One Girl at Three Ages

Ages (Years)	10	13	16
Height (in dm)	14	16	17
Mass (in kg)	31	52	65
Surface Area (in dm^2)	110	150	175
Volume (in dm^3)	41	67	79

Data adapted from *Arithmetic Teacher*,

16. Suppose that you want to draw a block that can be used to model the volume of the girl at age 13. Find the width and depth measurements of the block.

15. Estimates will vary. An easy unit of measure to use is cubes of 1 dm^3 (10 cm × 10 cm × 10 cm). The volume of the body in the picture on Student Book page 26 is 38 dm^3. The person is only 140 cm high and 10 cm thick.

16. Answers will vary. Sample strategies:

If the volume of the girl is 67 dm^3 and her height is 16 dm, then the base of the block could be $\frac{67}{16}$, or about 4 dm^2. So the width and depth would be about 2 dm × 2 dm (or 4 dm × 1 dm).

If a student is 157 cm tall, then because the model is 14 blocks high, each block would be $157 \div 14 \approx 11$ cm high or 11 cm on a side. Each cube would then have a volume of $11 \times 11 \times 11 = 1{,}331$ cm^3. There are 38 cubes in the model shown, so the volume of the shape would be $38 \times 1{,}331 = 50{,}578$ cm^3, which is the same as 50.5 dm^3.

The overall dimensions might be 16 dm × 2 dm × 2 dm or 16 dm × 4 dm × 4 dm.

Materials centimeter tape measures (one per group of students)

Overview Students estimate their body volume by modeling their body with cubes. They also find dimensions of a single block that can be used to model the given volume of a 13-year-old girl.

About the Mathematics Often in mathematics, more complicated problems are solved by looking at simpler cases, or situations that are "idealized" in some way. In the situation on Student Book page 26, a body modeled with cubes or rectangular blocks simplifies the problem of finding the volume of a person's body.

Planning Students may work on problems **15** and **16** in small groups. Discuss students' answers in class.

Comments about the Problems

15. It is not necessary for each student to complete the activity using their personal measurements. One person in each group could volunteer the necessary measurements.

16. Some students may not be familiar with the term *mass.* While technically it is a measure of an object's resistance to acceleration, it is probably sufficient to tell students that mass is the "stuff" that they are made of. Mass is the scientific term for what we know as "weight."

If a student is 157 cm tall, then, because the model is 14 blocks high, each block would be $157 \div 14 \approx 11$ cm high or 11 cm on a side. Each cube would then have a volume of $11 \times 11 \times 11 = 1{,}331$ cm^3. There are 38 cubes in the model shown, so the volume of the shape would be $38 \times 1{,}331 = 50{,}578$ cm^3, which is the same as 50.5 dm^3.

Other Measures for Volume

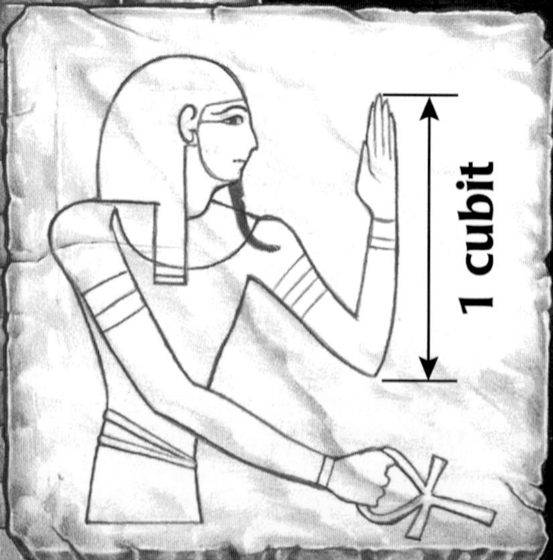

In the past, units of measure for volume were sometimes related to body sizes. It was more common, however, to use measurement tools like the cup as standards for measuring volume.

The ancient Egyptians had the first unit of measure for length: the cubit. The *cubit* is the distance from a person's elbow to the tip of the extended middle finger.

The Egyptians used a standard length called the *royal cubit,* which is 52.4 centimeters, or 20.62 inches.

In the ancient Egyptian system, the volume unit of measure is the *cubic cubit*.

17. a. Use your two arms to get a rough idea of how big a cubic cubit is.

b. About how many "royal" cubic cubits are there in 1 cubic meter?

The cubic fathom is a unit of measure used in the 1800s in Europe to measure volumes of firewood. (Remember that a fathom is the distance between your two outstretched arms.)

18. a. Estimate the amount of firewood that can fit into your cubic fathom.

b. What else can you measure with a cubic fathom? with a cubic cubit?

In the United States, firewood is measured in *cords.* A *face cord* is 4 feet high, 8 feet long, and 16 inches deep. A face cord usually fits in the back of a pickup truck. A *full cord* is 4 feet high, 8 feet long, and 4 feet deep.

19. How much more firewood is in a full cord than in a face cord?

Solutions and Samples
of student work

17. a. 50 cm^3. Arm positions may vary. One possibility is to bend one arm so that the forearm is horizontal and parallel to the chest. The other arm should be bent so that the fingers are pointing up and the elbow is in contact with the fingers of the other hand.

b. There are about 8 cubic cubits in 1 cubic meter. If 1 cubit = 50 cm, then 1 m $\approx$ 2 cubits and all sides are 2 cubits (2 cubits × 2 cubits × 2 cubits = 8 cubic cubits).

18. a. About 147 logs will fit into a person's cubic fathom that is 1.5 meters tall. Estimates will vary. Sample student response:

My height is 1.5 meters, so my cubic fathom is $(1.5\text{ m})^3 = 3.375\text{ m}^3$.

1.5 m deep is $\approx$ 3 logs
1.5 m high is $\approx$ 7 logs
1.5 m long is $\approx$ 7 logs
So together there would be 3 × 7 × 7 = 147 logs.

b. Answers will vary. Sample responses:

- cubic fathom: space in a car, bus, hallway, or classroom.
- cubic cubit: a laundry basket, boxes, cupboards, or a bushel basket.

19. There is three times more wood in a full cord than in a face cord. Students should reason that the face cord is 16 inches deep, while the full cord is 48 inches deep, so the full cord has three times as much wood.

Face cord: 4 ft × 8 ft × $1\frac{1}{3}$ ft = $42\frac{2}{3}$ ft^3.
Full cord: 4 ft × 8 ft × 4 ft = 128 ft^3.

Since the ratio of the volumes is 3:1, there is three times as much wood in the full cord.

Hints and Comments

Overview Students investigate historical measures of volume related to the body. They compare the measures *cubic cubit* and *cubic fathom* to cubic meters. Then they compare the volume of a face cord of firewood with that of a full cord.

About the Mathematics Doubling the dimensions (length and width and height) of an object does not result in a volume that is twice as big: the volume is enlarged by a factor of eight.

Planning Students may work on problems **17–19** in small groups. Discuss students' solutions and strategies for each problem.

Comments about the Problems

17. a. Students need not calculate an answer; they are merely getting an idea of size. They may, however, want to make a sketch and indicate dimensions in centimeters.

b. You may want to tell students to first find out how many cubits are in 1 meter before determining the volume of 1 cubic cubit in cubit meters.

18. a. Students responses may be in logs or in volume measures. If students are able to find one type of measure easily, they might be asked to find the other.

You may want students to outline the base of a cubic fathom on the ground to see how big it would be (once they figure in the height).

19. Encourage students to draw sketches indicating the dimensions of a full cord and a face cord. The sketches may show them that it is not necessary to calculate both volumes to compare them since only the depths are different.

Summary

Volume is important for measuring the sizes of such objects as a box, a drink, or even your hand.

In this section, you measured volumes in two ways: with liquid units of measure (such as liters) and with solid units of measure (such as cubic centimeters).

You also measured the following:

- the volume of an object when you wanted to know how much space it takes up or how much it holds.
- the surface area or area of an object when you wanted to know what it takes to cover it.
- the length of an object when you wanted to know how long or how tall it is.

Summary Questions

20. **a.** When would it be useful to know the volume of an object?

b. Pick an object and estimate its volume. You may use any units and any measuring method you wish.

21. Linda knows that 1 liter equals "one cubic something." She doesn't remember whether it is 1 cubic centimeter, 1 cubic decimeter, or 1 cubic meter. Explain to Linda which unit of measure equals 1 liter.

Solutions and Samples
of student work

20. a. Answers will vary. Sample response:

You might need to know an object's volume for packing, shipping, or pricing.

b. Answers will vary, depending on the object chosen. Students may choose the objects and units listed below:

- volume of the classroom: measure in cubic fathoms, cubic meters
- volume of a soda can: measure with liquid (in milliliters) or sand (in cubic centimeters)
- volume of a book: measure in cubic centimeters

21. A liter is equal to 1 cubic decimeter. Explanations will vary. Students may make a net of 1 cm^3 and point out that it is much too small. They may mention that 1 m^3 is used to measure very large objects, such as cars. If students use the example of a liter bottle, it is easy to show with a centimeter cube and a marker that 1 dm^3 is the right choice.

Hints and Comments

Materials metric tape measures and beakers of various sizes (several per class)

Overview Students read the section Summary and answer two problems involving volume.

Planning Have students read and discuss the Summary. They may work on problems **20** and **21** individually. These problems may be used for assessment. Discuss students' answers in class.

Comments about the Problems

20. Informal Assessment This problem assesses students' ability to estimate lengths, areas, and volumes visually using appropriate units; to find (compute, measure, and draw) lengths, areas, and volumes using standard and nonstandard units; and to recognize that different units of measurement and different measurement systems exist. It also assesses their ability to use the relationships between length, area, and volume; to use formulas and rules of thumb to estimate and calculate length, area, and volume; and to select and use appropriate measurement units.

21. Informal Assessment This problem assesses students' ability to use simple relationships between and within measurement systems.

Work Students Do

Students use the context of ergonomics (the science concerned with designing safe and comfortable objects, such as furniture, to learn about body angles. They explore the arc of movement of the wrist by marking angles on a piece of paper. They examine how angles can be used to design a keyboard and a computer monitor. Finally, students see how angles are taken into account when designing chairs.

Goals

Students will:

- estimate lengths, areas, volumes, and angles using appropriate units;
- find (compute, measure, and draw) lengths, areas, volumes, and angles using standard and nonstandard units;
- use geometric models to solve problems.

Pacing

- approximately four 45-minute class sessions

About the Mathematics

Angles in this section are measured in degrees (out of 360° around a circle). Measuring angles on the human body gives students the opportunity to think more about angles in more than one plane. For example, the elbow joint can move in the same plane of the body as well as horizontally.

In the context of easy chairs and the angles that their seats and backrests make to the horizontal plane (Student Book page 33), students measure the angles for each chair from the same zero degree line. (Although the two chairs pictured on the page will have different zero lines.) See the illustration below. Students will learn in later units that in the coordinate grid system, angles are always measured from the positive *x*-axis, in a counterclockwise direction.

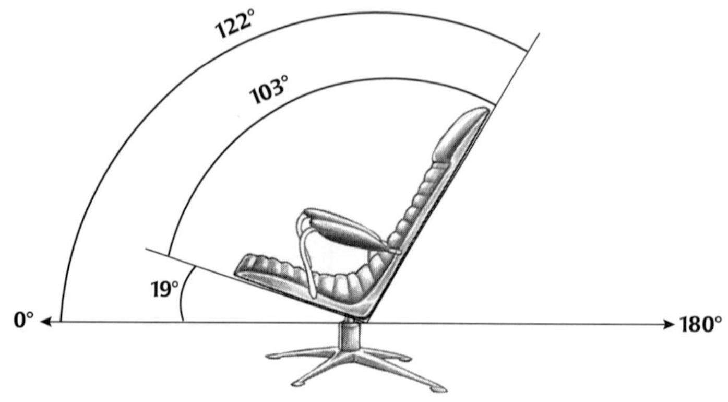

Materials

- dictionary or encyclopedia, page 69 of the Teacher Guide (one per class)
- $8\frac{1}{2}''$ × 11″ paper, pages 71 and 73 of the Teacher Guide (two blank sheets per student)
- compass cards from Transparency Master 1 or protractors, pages 73, 75, 77, and 79 of the Teacher Guide (one per student or group of students)
- string, page 79 of the Teacher Guide, optional (one piece about 1.5 meters long per group of students)
- pens, page 79 of the Teacher Guide, optional (one per group of students)
- drawing chalk or sheets of paper about 2 square meters, page 79 of the Teacher Guide, optional (one per group of students)

Planning Instruction

To start this section, you may want to have a class discussion about ergonomics. You might discuss the design features of the desks and chairs in your classroom or bring in an uncomfortable chair and ask students why it is uncomfortable and how they might improve upon it. In this discussion, students might mention angles, in addition to statements such as "it needs a bigger seat," and so on.

Students should have experience with obtuse, right, and acute angles introduced in the unit *Figuring All the Angles*. You may ask them to classify the angles that they encounter in this section in this way.

Students may work on problems 1, 7, 8, and 10 in small groups. Problems 2 and 4a may be done as a class. Students may work on problems 3, 4b-6, and 9 individually.

There are no optional problems in this section.

Homework

Problems 5 and 6 (page 74 of the Teacher Guide), and 9 (Page 76 of the Teacher Guide) may be assigned as homework. The Extension activity (page 73 of the Teacher Guide) and the Bringing Math Home activities (pages 69 and 77 of the Teacher Guide) may also be assigned as homework. After students complete Section D, you may assign appropriate activities from the Try This! section, located on pages 35–37 in the *Made to Measure* Student Book, as homework. The Try This! activities reinforce the main mathematical concepts covered in this section.

Planning Assessment

- Problem 3 can be used to informally assess students' ability to estimate angles using appropriate units and to find (compute, measure, and draw) angles using standard units.
- Problem 9 can be used to informally assess students' ability to find (compute, measure, and draw) angles.
- Problem 10 can be used to informally assess students' ability to estimate angles using appropriate units; to find (compute, measure, and draw) angles; and to use geometric models to solve problems.

D. ANGLES

INTRODUCTION

When designing furniture, cars, computers, and other items, it is sometimes necessary to know the angles that a person can make with his or her arms, ankles, wrists, and head.

1. For what objects might it be important to keep angles in mind?

Ergonomics is important for designing work environments. Designers use ergonomics to determine the placement of office equipment and the size and dimensions of furniture that will create safe, efficient working environments for people. Designers also consider ergonomics when creating new buses and trains for commuters, for example, by placing controls close to the driver.

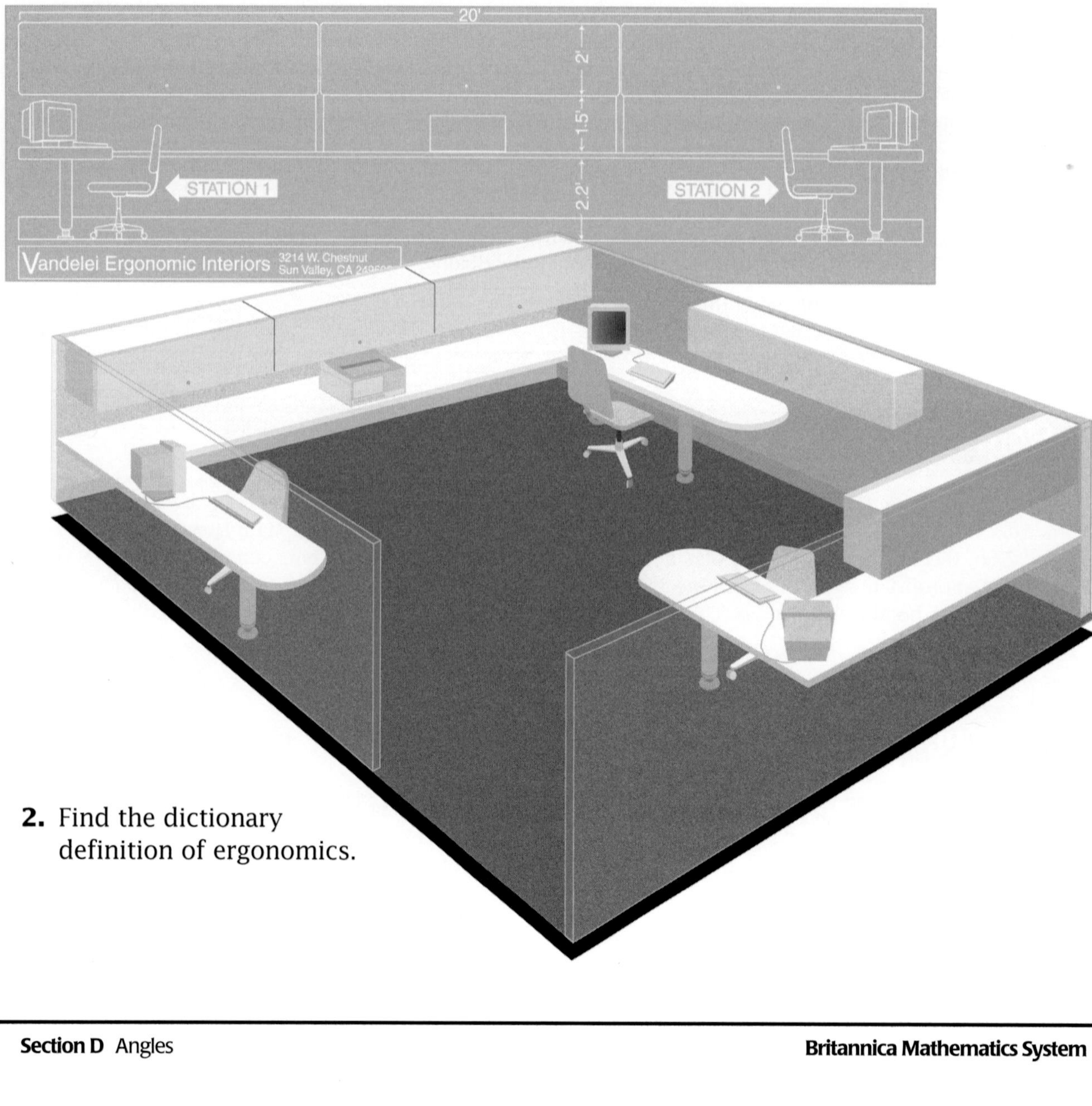

2. Find the dictionary definition of ergonomics.

Solutions and Samples
of student work

1. Answers will vary. Sample responses:
   - a computer keyboard
   - a car (the positions of the steering wheel, the brake pedal, and so on)
   - a piano (or other musical instrument)
   - a chair
   - a bike
2. *Ergonomics* is the study of people and their work environments. Some of the goals of ergonomics are the following:
   - to reduce stress on a person's body as caused by interaction with and use of objects;
   - to design equipment and machines for safe and efficient use;
   - to construct work stations to ensure correct body posture; and
   - to adjust lighting, air-conditioning, and noise levels for optimal working conditions.

Hints and Comments

Materials dictionary or encyclopedia (one per class)

Overview Students learn the meaning of the word ergonomics; how it involves learning about people's range of movement; and how it impacts the design of such things as chairs and computer keyboards.

Planning Problem **1** may be answered in small groups or individually; problem **2** can be done as a class.

Comments about the Problems

1. After investigating an uncomfortable chair, it is very likely that students will mention all sorts of furniture. Encourage them to think of other objects as well, such as musical instruments, vehicles, and so on.

2. You may want to ask one student to look up the word and then share the definition with the class. Also, you may want to have students consult other reference books. Information in encyclopedias may be found under *systems engineering.* Another good reference is *Fitting the Task to the Man,* by Etienne Grandjean.

Bringing Math Home Ergonomics does not need to be restricted to the workplace. Students may look around at home to see for which objects it is important to keep comfortable designs in mind and talk about this issue with their families. They may look at chairs, countertops, tables, doors, bathtubs, toilets, sinks, and so on. Have them focus on the height and placement of objects, paying particular attention to angles.

Activity

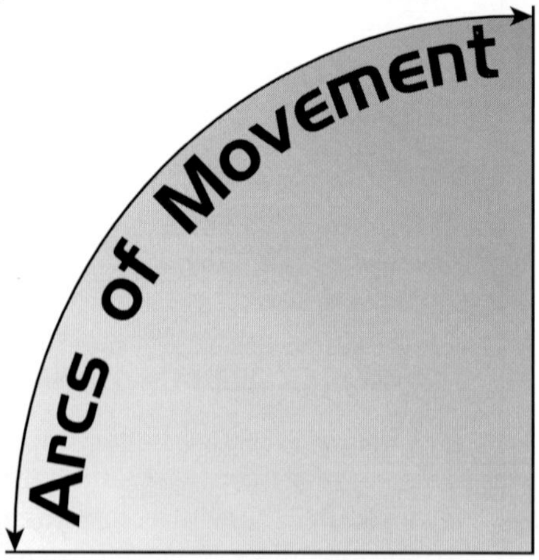

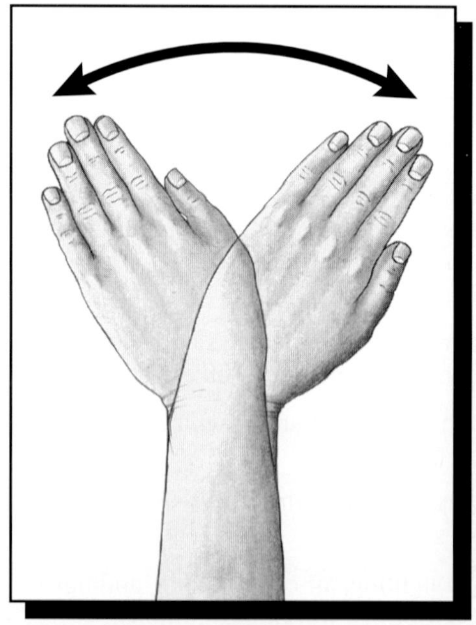

On the right is the "arc of movement" of someone's left hand. The picture shows how far to the left and right the wrist can bend.

You can measure the arc of movement of your own hands.

i. Draw a small × on the bottom of a sheet of paper.

ii. Lay your forearm flat, with your left hand and wrist on the sheet of paper. Put the middle of your wrist over the small ×.

iii. Put a mark on the paper at the top of your middle finger.

iv. With your arm and hand flat on the table, bend your wrist to the right as far as you can without moving your arm. Put another mark on the paper at the top of your middle finger.

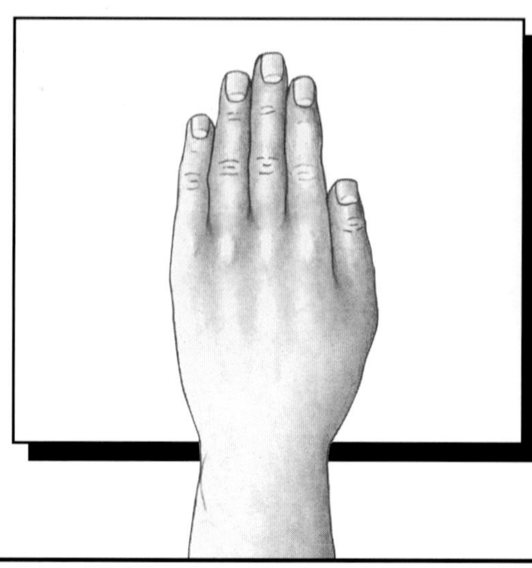

Solutions and Samples
of student work

Hints and Comments

Materials $8\frac{1}{2}''$ × 11″ paper (one sheet per student)

Overview Students draw the arc of movement of their left hand and measure the angle formed.

Planning Students may work on steps **i–vi** individually. It is important that the forearm stays fixed on the table. Students may mark their wrist with an × as well to make sure that it stays on the × on the sheet of paper.

The three lines students draw on their paper should be equally long, since all three represent the length of their middle finger.

v. Bend your wrist to the left as far as you can without moving your arm. Mark the location of the top of your middle finger.

vi. Draw lines from each of the three finger marks to the small × (the middle of your wrist).

3. a. Use your protractor to measure the angles that your left hand can bend to the right and to the left.

b. What do you think the angles will be for your right hand? Measure the arc of movement of your right hand following the same instructions on page 30 and above.

4. a. Make a table of the right hand angle measurements for all of the students in your class. Describe the results.

b. Why might someone be interested in studying the motion of the wrist joint?

3. a. The measures will be about 33° on the thumb side and 45° on the little finger side.

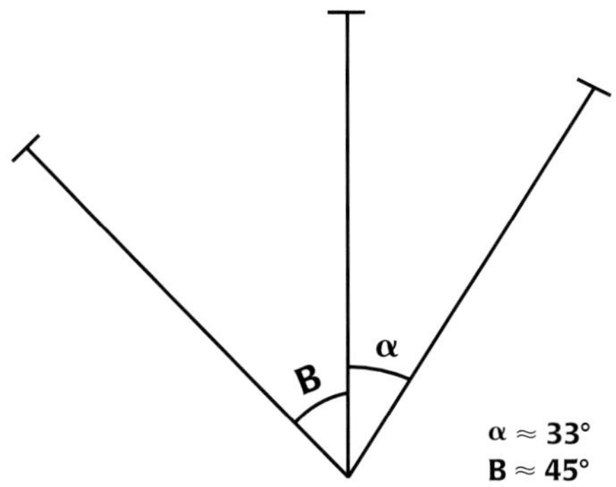

b. The right-hand angles will mirror the left-hand angles. The smaller angle will occur on the thumb side.

4. a. Answers will vary. In one pilot class, the mean angle measure on the thumb side was 32° and on the little finger side was 40.58° The range of angle measures on the thumb side was 24° to 40°. The range on the little finger side was 30° to 50°. The mode or median can also be used to describe the results.

b. Answers will vary. Students should mention that wrist movement data is important to designers of objects like computer keyboards.

Materials compass cards from Transparency Master 1 or protractors (one per student); paper (one blank sheet per student)

Overview Students measure the angles that describe the range of movement of their left hands using the drawing they made on the previous page. They then measure the range for their right hands. They compare the right-hand angle measurements for all students in the class, and think about why the motion of the wrist joint might be important in ergonomics.

Planning Students may work individually on problem **3.** Problem **4a** is a whole-class activity. Students may work on problem **4b** individually. Problem **3** may be used for assessment.

Comments about the Problems

3. Informal Assessment This problem assesses students' ability to estimate angles using appropriate units and to find (compute, measure, and draw) angles in standard units. Students must decide how to place their compass cards or protractors on their drawings.

4. a. Students might represent the data in a histogram or use one-number statistics (mean, median, mode) to summarize the data.

Extension You may wish to have students investigate whether the angles of their dominant hand are bigger than the angles of their other hand or whether people with bigger hands make bigger angles.

When typing on a computer keyboard, your hands should rest on the keys. In order to reach all of the keys, your hands must bend sideways. If you do a lot of typing, the position in which you hold your hands on the keyboard can cause physical discomfort in your hands, wrists, or forearms.

Studies on ergonomic keyboard design have found that most people prefer a split keyboard. A *split keyboard* is divided into two parts: the left part ends with the keys T-G-B, and the right part starts with keys Y-H-N.

On a split keyboard, the two parts make an angle of about 25°, that is, the lines drawn through T-G-B and Y-H-N make an angle of 25°. The distance between the two parts—or between the keys G and H—should be about 95 millimeters.

5. a. Make a drawing of a split keyboard. You do not have to draw all of the keys, only an outline of the two parts.

b. What is one advantage to using a split keyboard?

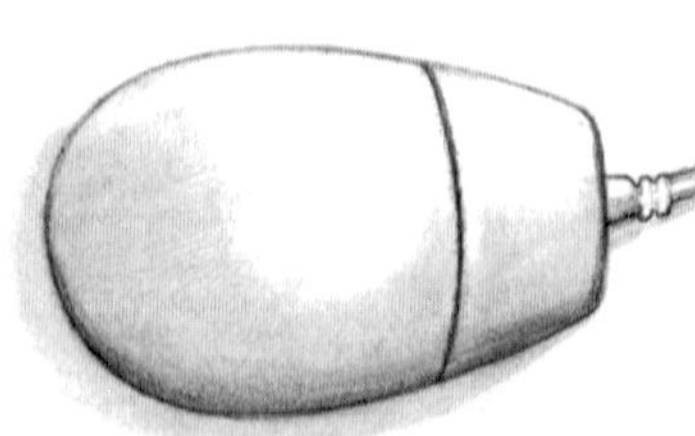

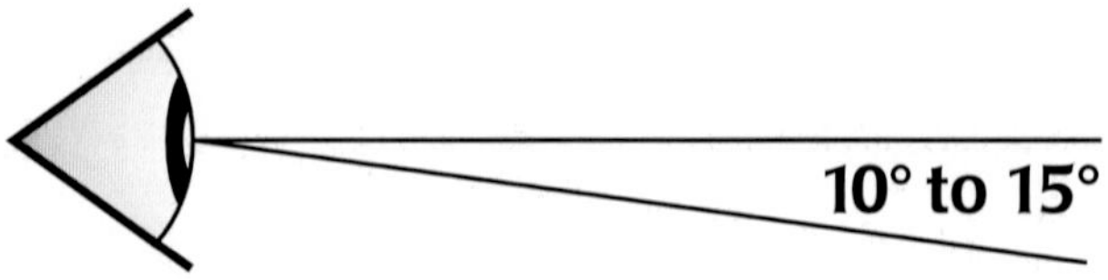

Another important angle in design involves your *line of sight*. If you look straight ahead, your line of sight follows a horizontal line. Your normal line of sight is typically 10° to 15° below the horizontal.

In the best ergonomic computer design, the screen should not be lower than the sight line of 15° below the horizontal.

6. Draw a picture of the situation described above.

5. a. Drawings will vary in their level of detail, but the angle and distance measurements given should be represented in the sketch. Sample student drawing:

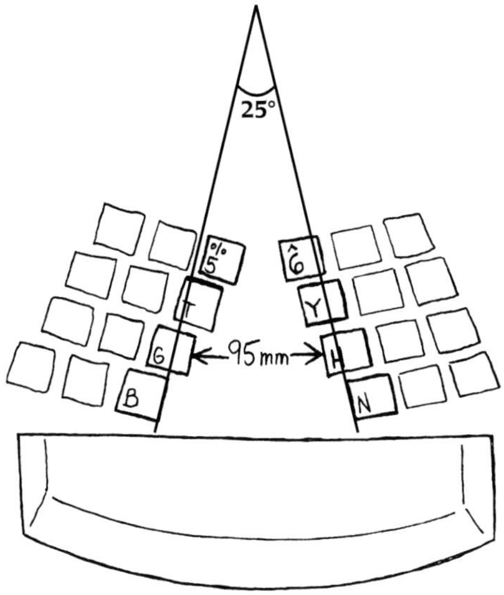

Students can draw one line, measure the distance (95 millimeters), plot a point at that distance, and then use a protractor to find the line through the point that makes an angle of 25° with the first line.

b. Answers will vary. Basically, the wrists do not need to bend as much with a split keyboard, which makes typing more comfortable.

6. Drawings will vary. Sample student drawing:

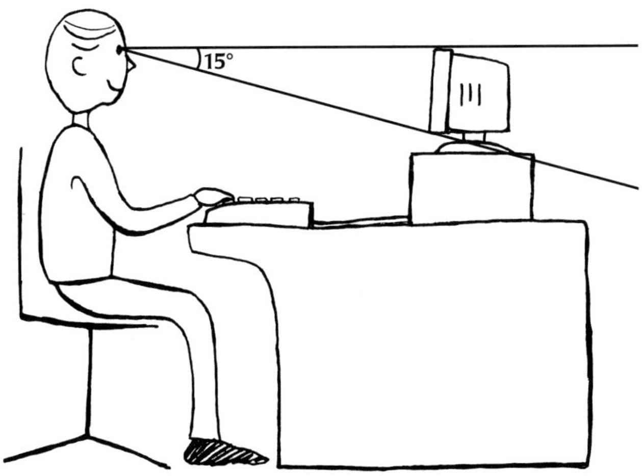

Materials compass cards from Transparency Master 1 or protractors (one per student)

Overview Students read a description of a split keyboard and then draw one. They think about the advantages of this kind of keyboard. They also draw lines of sight that make a given angle, which is important for the ergonomic design and placement of computer screens.

Planning Before students begin work on problem **5,** make sure they understand what they are expected to do. Students may work on problems **5** and **6** individually. Have students then share their work in class. These problems may be assigned as homework.

Comments about the Problems

5–6. Homework These problems may be assigned as homework.

5. a. Students might make a scale drawing of a split keyboard, although it is possible to draw the middle of the keyboard to full scale.

b. To answer this problem, students may use their data on the hand arcs from Student Book page 31.

6. The purpose of this question is to have students draw the angle (by first drawing the line to indicate the horizontal plane). Tell students that the drawing does not need to be a work of art. It should include a person, a computer, and lines of sight.

You can also use ergonomics to design an easy chair. Some typical recommendations for the design of an easy chair follow.

- The seat should be tilted backward so that you will not slide off. A tilt of 14 to 24 degrees to the horizontal is recommended.
- The angle between the backrest and the seat should be from 105 to 110 degrees, and the angle between the backrest and the horizontal should be from 110 to 130 degrees.

Shown below are two different easy chairs.

7. Check to see if these chairs fit the above recommendations.

8. Based on the above recommendations, design your own easy chair. Draw a side view of your chair on an appropriate scale.

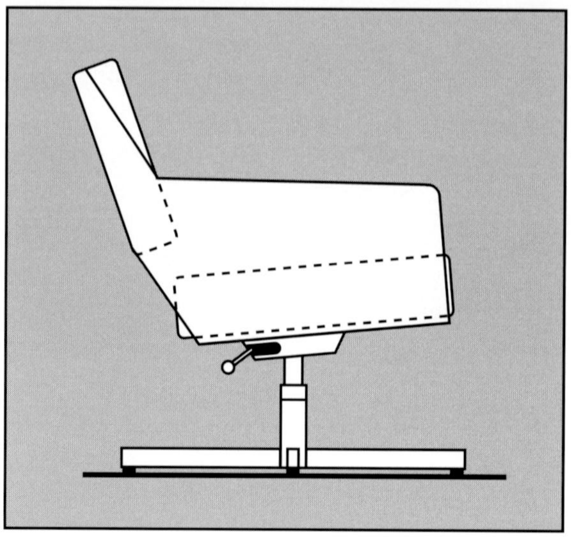

Shown on the left is the side view of an office chair.

9. **a.** Measure the angles the seat and the backrest make to the horizontal plane. Also measure the angle of the backrest to the seat.

   b. Compare your findings to the above recommendations for easy chairs. What conclusion can you make?

Solutions and Samples
of student work

7. Neither chair fits all recommendations.

   The chair pictured on the left has a seat that is tilted backwards about 9° from the horizontal plane, which is too little. The angle between the seat and the backrest for the left chair is about 103°, which is acceptable. To the horizontal plane, the backrest is tilted 9° + 103° = 112°. This is also acceptable. The left chair, therefore, fits two of the three recommendations.

   For the right chair, the seat is tilted backwards about 19° from the horizontal plane, which is acceptable. The angle between the seat and the backrest is about 103°, which is not quite enough. To the horizontal plane, the backrest is tilted 19° + 103° = 122°. This is acceptable. The right chair, therefore, fits two of the three recommendations.

8. Designs will vary. Students' chairs should fit the recommendations and be drawn to scale.

9. **a.** The seat is about 5° to the horizontal plane, while the backrest is about 110° to the horizontal. The angle of the backrest to the seat is about 105° (the difference between the two angles).

   b. Answers will vary, but students should note that the seat barely fits the recommendations, which is not surprising since an office chair is not an easy chair.

Hints and Comments

Materials compass cards from Transparency Master 1 or protractors (one per student)

Overview Students learn about recommendations for the design of an easy chair and then judge two easy chairs and an office chair on the basis of these recommendations. They also design their own easy chair.

Planning Have students read the text at the top of the page and work on problems **7** and **8** in small groups. Discuss their answers to problems **7–9** in class. Students may work on problem **9** individually. It may be assigned as homework or used for assessment.

Comments about the Problems

7. Students may want to trace the outlines of both chairs and draw a horizontal line before measuring the angles involved with a compass card or protractor.

   Remind students that when they measure the angle of the backrest to horizontal, they must measure from the same zero line they used to measure the seat angle (although the zero line will change from chair to chair).

   Some students may notice that the recommended minimums for the seat to horizontal and for the backrest to the seat do not add up to the minimum for the backrest to horizontal. The same holds true for the maximums. Each angle was evidently tested separately for comfort.

9. **Informal Assessment** This problem assesses students' ability to find (compute, measure, and draw) angles.

   a. Again, students may trace the outline of the seat and the backrest and draw a horizontal line before measuring angles.

   b. Office chairs are a bit more upright than easy chairs.

Bringing Math Home Students might use the given recommendations to judge easy chairs at home. They can rate the comfort of these chairs or ask family members to rate them. Students can then write reports about their findings.

Summary

In this section, you investigated the movement for both the right and left hand. You also explored how the placement of a computer screen and the angles between the seat and the backrest of an easy chair can affect your comfort.

Summary Questions

The following text is taken from a book on ergonomics, *Fitting the Task to the Man*, by Etienne Grandjean.

"The arm can rotate through an angle of 250° about its axis...of which a half-circle (180°) lies in front of the body, and a further 70° or thereabouts, backwards."

10. **a.** Read the text carefully and explain what it means.

b. Check to see if your arm can make an angle of 70° backward. First think of a way to measure this angle.

10. a. Explanations will vary. The text refers to rotating the arm in the horizontal plane, moving it extended from the front of the body to the back. It can rotate through 180° in front of the body and another 70° behind it.

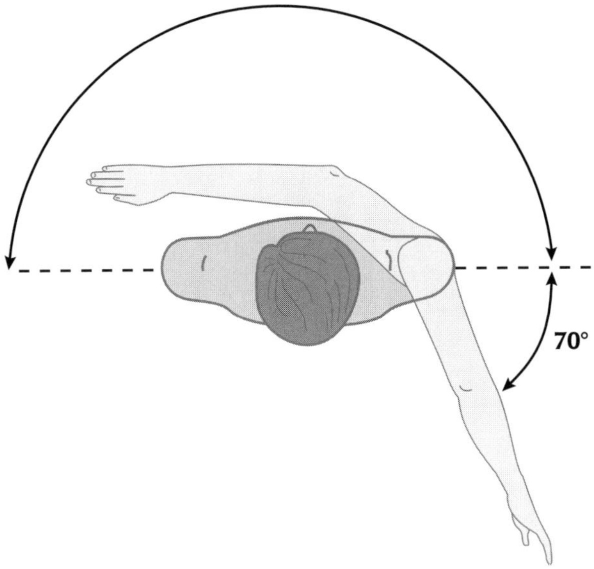

b. Student data will vary. A compass card or protractor can be positioned so that the center is at the joint of the arm and the torso. The angle measure can then be read after a student has moved his or her arm. Two people are needed to do this. Each student can lay his or her arm on a table on a large sheet of paper. Another student can draw lines starting at the shoulder for the positions of the arm and then measure the angle.

Materials compass cards from Transparency Master 1 or protractors (one per group); string, optional (one piece about 1.5 meters long per group); pens, optional (one per group); drawing chalk or a sheet of paper of about 2 square meters, optional (one per group)

Overview Students read and discuss the Summary and text taken from a book on ergonomics, which focuses on the angle of rotation of a person's arm. They check the statement in the text by measuring the rotation angle of their arm.

Planning Have students read the Summary and study the text from *Fitting the Task to the Man.* Students may work on problem **10** in small groups in class. This problem may also be used as a group assessment.

Comments about the Problems

10. Informal Assessment This problem assesses students' ability to estimate angles using appropriate units; to find (compute, measure, and draw) angles; and to use geometric models to solve problems.

a. You may want students to make a sketch showing what is described in the text. Such a drawing will help them to see how to complete part **b.**

Ask students to explain why the 145° is fixed for every person (because the arm is blocked by the body).

b. Students should think of a good way to measure this angle. It works best to have them carry out the measuring in groups. You may ask students to write a short report about their measuring method and their findings, including sketches that show what they measured and how.

Students may come up with different measuring methods. Some students may draw a huge compass card on the floor (with refinement up to 10°) and have one student stand on it with the shoulder right above the center, outstretch the arm, and use a plumb line attached to the hand to measure the angle between the side position and the back position.

Assessment Overview

Students work on five assessment problems and one end-of-unit project that you can use to collect additional information about what each student knows about using standard and nonstandard measurement units to estimate and find lengths, areas, volumes, and angles.

Goals	Assessment Opportunities
• estimate lengths, areas, volumes, and angles using appropriate units	Problems 1a, 1c, 3b, and 5c Made to Measure Project
• find (compute, measure, and draw) lengths, areas, volumes, and angles using standard and nonstandard units	Problems 1a, 2, 3a, 4a, 5a and b Made to Measure Project
• use simple relationships between and within measurement systems	Problems 1d, 2, and 5b Made to Measure Project
• recognize that different units of measurement and different measurement systems exist	Problems 1c and 2 Made to Measure Project
• recognize the importance of standard (measurement) units	Problems 1c and 2 Made to Measure Project
• understand and use the relationships between length, area, volume, and angle	Problems 1e, 3b, 3c, and 4b Made to Measure Project
• use geometric models to solve problems	Problems 3a–c Made to Measure Project
• use formulas and rules of thumb to estimate and calculate length, area, and volume	Problems 1e, 3b and 3c Made to Measure Project
• develop and use a frame of reference for measurement to solve problems	Problems 1f, 2, and 5c Made to Measure Project
• select and use appropriate measurement units	Problems 1b, 1e, and 5b Made to Measure Project

Pacing

When combined, the five assessment problems and one project will take approximately two 45-minute class sessions. See the Planning Assessment section for further suggestions as to how you might use the assessment activities.

About the Mathematics

The five assessment problems and end-of-unit project evaluate the major goals of the *Made to Measure* unit. Refer to the Goals and Assessment Opportunities sections above for information regarding the specific goals assessed in each assessment activity. Students may use different strategies to solve each problem. Their choice of strategies may indicate their level of comprehension of the problem. Consider how well students' strategies address the problem, as well as how successful students are at applying their strategies in the problem-solving process.

The Made to Measure Project involves taking measurements of students' own bodies using appropriate units and describing the relationships between the measurements found for different body parts.

Materials

- Assessments, pages 94–97 of the Teacher Guide (one of each per student)
- centimeter rulers, pages 83, 87, and 89 of the Teacher Guide (one per student)
- inch rulers, page 89 of the Teacher Guide (one per student)
- compass cards from Transparency Master 1 or protractors, pages 87 and 89 of the Teacher Guide (one per student)
- grid paper, pages 87 and 89 of the Teacher Guide, optional (one sheet per student)
- calculators, pages 83, 85, 87, and 89 of the Teacher Guide (one per student)

Planning Assessment

You may want students to work on the end-of-unit assessment problems individually to evaluate each student's understanding and abilities. Make sure that you allow enough time for students to complete each activity. Students are free to solve each problem in their own way. They may choose to use any of the models introduced and developed in this unit to solve problems that do not call for a specific model.

Since the five assessment problems assess the same goals as the Made to Measure Project, you can use either assessment tool to evaluate your students. Both assessment options each take approximately one class period to complete. Some of the five assessment problems can be used at the end of a section in which the related goals are introduced. You might want to assign the Made to Measure Project toward the end of the unit as a long-range homework assignment, or you may decide to have students complete this activity in class during one class session.

Scoring

Answers are provided for all assessment problems. The method of scoring the problems depends on the types of questions in each assessment. Most questions require students to explain their reasoning or justify their answers. For these questions, the reasoning used by the students in solving the problems, as well as the correctness of the answers, should be considered as part of your grading scheme. A holistic scoring plan can be used to evaluate an entire task. For example, after reviewing a student's work, you may assign a key word, such as *emerging, developing, accomplishing,* or *exceeding,* to describe their mathematical problem-solving, reasoning, and communication.

On other tasks, it may be more appropriate to assign point values for each response. Student progress toward the goals of the unit should also be considered. Descriptive statements that include details of a student's solution to an assessment activity can be recorded. These statements will provide insight into a student's progress toward a specific goal of the unit. Descriptive statements are often more informative than recording only a score and can be used to document student growth in mathematics over time.

See page 89 of the Teacher Guide for suggestions as to how you might score students' measurement findings for the Made to Measure Project.

MADE TO MEASURE PROBLEMS

Use additional paper as needed.

1. **a.** Use one or more historical measurement units that are related to the human body to find the length and width of this sheet of paper.

   b. Explain why you chose the measurement unit(s) you did.

   c. Explain why it is not likely that everyone in your class got the same answer for part **a.**

When measured in centimeters, this paper is about 21 centimeters wide and 28 centimeters long.

   d. Write two statements that show the relationship between the historical measurement unit(s) you used in part **a** and centimeters.

   e. Estimate the area of this paper in historical measurement units *and* in square centimeters.

Suppose you make a cube out of this sheet of paper by using a net similar to the one shown below. Suppose the net is large enough to just fit this sheet of paper.

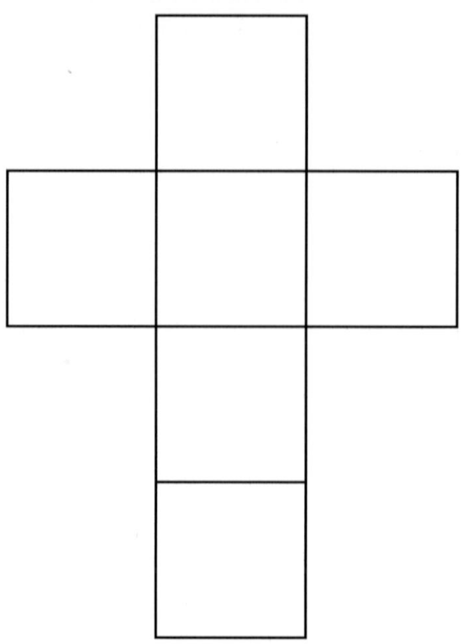

   f. Will this cube hold more or less than one liter? Explain your answer.

Solutions and Samples
of student work

1. **a.** Answers will vary. Students may use the hand span or thumb measurement units to determine the length and width of the paper. Others may use the length of their thumb or the width of their hand. Accept any reasonable answers.

   b. Explanations will vary. Sample explanation:

   I measured the dimensions of the paper in hand spans and thumbs because the other historical measurement units are all too large. One foot or one yard would be longer than the length or width of the paper.

   c. Explanations will vary. Sample explanations:

   Everyone's answers will differ since different studies may have used different measurement units.

   Since every student has different-sized hands and thumbs, their answers will be different.

   It is difficult to measure the dimensions of the paper accurately using historical units. Even if everyone used the same measurement unit, they would probably come up with different answers.

   d. Answers will vary. Sample responses:

   One hand span is equal to about 15 centimeters. One thumb is equal to about 2 centimeters.

   e. Answers will vary for the area of the paper expressed in historical units. Sample responses:

   The area of the paper is about 5 handprints.

   The area of the paper is about 2 square hand spans.

   The area of the paper in square centimeters is 588 cm^2 (21 cm × 28 cm = 588 cm^2). An estimate such as 600 cm^2 (20 cm × 30 cm = 600 cm^2) is also acceptable.

   f. The cube will hold less than one liter. Explanations will vary. Sample explanation:

   By sketching the net on a sheet of paper, I found its dimensions to be 7 cm by 7 cm by 7 cm. So, its volume is 343 cm^3. To hold one liter, its dimensions must be 10 cm by 10 cm by 10 cm, which is 1,000 cm^3 or 1 dm^3.

Hints and Comments

Materials Made to Measure Problems assessment, page 94 of the Teacher Guide (one per student); centimeter rulers (one per student); calculators (one per student)

Overview Students use nonstandard and standard measurement units to estimate and find the length, width, and area of a sheet of paper. They also describe the relationships between the nonstandard units used and centimeters.

About the Mathematics This assessment problem evaluates students' ability to estimate and find (compute, measure, and draw) lengths, areas, volumes, and angles using standard and nonstandard units; to recognize that different units of measurement and different measurement systems exist; and to recognize the importance of standard (measurement) units.

Planning You may want students to work individually on problem **1.** This problem can be assigned after students finish Section C or as an end-of-unit assessment.

Comments about the Problems

1. **a.** You may need to remind students about the historical nonstandard units used in Section A: thumb, hand span, foot, yard, pace, and fathom.

   d. Be sure that students use their measurements from part **a** to answer this question. For example, if students found the length of the paper to be 2 hand spans, they should note that 1 hand span is equal to about 14 centimeters. Answers such as "My hand span is less than 21 centimeters" are less accurate but should be considered correct.

   e. Again, be sure that students use their measurements from part **a** to estimate the area of the paper using nonstandard units. Students may use the unit handprints as a measure or some square measure (such as square hand spans or square thumb lengths). A student giving an answer like "one hand span times two hand spans" should receive only partial credit.

MADE TO MEASURE PROBLEMS

Use additional paper as needed.

In a survey, people were asked how far they commute daily from home to their place of work. Here are some of their responses:

Jody said, "I commute about three miles to get to work every day."

Carl said, "It takes me about 15 minutes to get to work."

Brenda said, "I live a little less than 5 kilometers from work."

Edward said, "It takes me about 10 minutes to get to work by car."

2. a. Of the four people mentioned above, who do you think lives the farthest from his or her place of work? Explain your reasoning.

b. Which two of these people live the same distance from their place of work? Explain.

Hannah has a dog named Fido. Shown below is a picture of Fido and some of Fido's body measurements.

- Body height: 36 centimeters
- Height of legs: 30 centimeters
- Body length: 44 centimeters
- Circumference around body: 45 centimeters
- Head length: 18 centimeters

Hannah has taken some measurements of Fido, which are shown above. Hannah wants to know what Fido's surface (skin) area is.

3. a. Describe at least two ways that Hannah could estimate Fido's surface area.

b. Estimate Fido's surface area. Explain how you made your estimation.

c. Estimate Fido's volume in cubic decimeters.

Solutions and Samples
of student work

2. a. Answers will vary. Accept answers that indicate Carl or Edward live the farthest from work. Explanations will vary. Sample explanation:

Since Jody travels 3 mi to work and Brenda travels 5 km to work, they both live about the same distance from work (1 mi is about 1.6 km, so 3 mi is about $1.6 \times 3 = 4.8$ or 5 km). A person could probably travel 3 mi or 5 km in about 5 minutes, so Jody and Brenda do not travel as far to work as Edward, who lives 10 minutes away from work. The problem does not say how Carl gets to work. If Carl drives 15 minutes to work, he lives farther from work than Edward, who travels 10 minutes to work. If Carl walks 15 minutes to work, he lives closer to work than Edward.

b. Brenda and Jody live the same distance from work. Explanations will vary. See the sample explanation in part **a.**

3. a. Estimation strategies will vary. Sample strategies:

- Cover Fido in newspapers. Then remove the newspaper and reallot the different pieces into a rectangle. Measure the length and width of the rectangle to estimate Fido's surface area.
- Multiply the dog's body length (44 cm) and width (the circumference around the belly [45 cm]) to estimate the dog's surface area.
- Think of Fido's body as a large block. Use the dimensions of the block to find its surface area.

b. Estimates will vary. Sample response using the given measurements in the diagram:

I imagined Fido's body as a block. The length of the block is about 60 cm (head length and body length), the width and the height of the block are both about 18 cm (half the height from the ground).

So, the surface area is about (4×60 cm $\times 18$ cm) + (2×18 cm $\times 18$ cm) = 4,320 cm^2 + 648 cm^2 = 4968 cm^2 or about 5,000 cm^2.

c. Estimates will vary. Sample response:

I imagined Fido's body as a block. I converted the dimensions of the block given in centimeters to decimeters:

60 cm = 6 dm; 18 cm = 1.8 dm; 18 cm = 1.8 dm

Then I multiplied to estimate the volume:

$6 \text{ dm} \times 1.8 \text{ dm} \times 1.8 \text{ dm} \approx 19.5 \text{ dm}^3$

Hints and Comments

Materials Made to Measure Problems assessment, page 95 of the Teacher Guide (one per student); calculators (one per student)

Overview Students analyze four survey responses that describe commute distances in terms of time units and standard units to determine which person has the longest commute to work. Student then use the given body measurements of a dog to estimate its surface area and volume.

About the Mathematics Problem **2** assesses students' ability to find (compute, measure, and draw) lengths using standard and nonstandard units; to use simple relationships between and within measurement systems; to recognize that different units of measurement and measurement systems exist; and to develop and use a frame of reference for measurement to solve problems.

Problem **3** assesses students' ability to find areas using standard and nonstandard units. It also assesses their ability to estimate areas using appropriate units; to use the relationships between length, area, and volume; to use geometric models to solve problems; and to use formulas and rules of thumb to estimate and calculate length and area.

Planning You may want students to work individually on these problems.

Comments about the Problems

2. Focus on students' reasoning for this problem rather than on their answers. Only responses that mention the fact that not enough information is given regarding Carl's commute should receive full credit. Students may use what they know about a person's average walking and driving speeds to estimate Carl's and Edward's distances. They will have to make explicit assumptions about whether Carl walks or drives to work and the impact that this has on their commute.

3. a. You may need to explain to students that some surface area estimation methods introduced in Section B, such as the "personal square method," cannot be used to find the surface area of a dog.

c. Remind students that the volume measurement should be expressed in cubic decimeters.

MADE TO MEASURE PROBLEMS Page 3 of 3

Use additional paper as needed.

The drawing below shows a persons ankle stretching as far up as it can go.

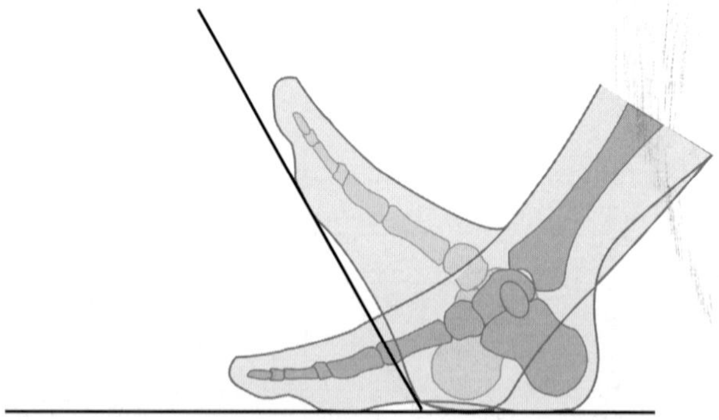

4. a. Measure the outer angle formed by the two positions of the ankle.

b. Which of the two positions looks the most comfortable for an ankle. Explain why.

"Climbing ladders is most convenient when the ladder is at an angle of 70° and the rungs are 260 millimeters apart." E. Grandjean

5. a. The picture on the right shows the side view of a wall. Draw a ladder against it that is at an angle of 70 degrees with the ground.

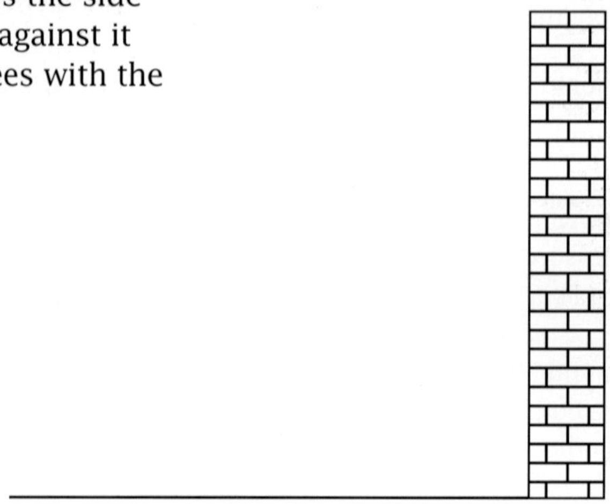

b. On the back of this paper, draw a ladder that is 2 meters long and 50 centimeters wide to scale. Remember to draw the rungs 260 millimeters apart.

c. Find an object in your classroom that is about as tall as the ladder you drew.

Solutions and Samples
of student work

4. a. The angle is 60°.

b. The most comfortable position for an ankle is when the foot is resting flat on the floor.

5. a. Drawings will vary, but the angle formed by the foot of the ladder and the ground or floor should be 70°. Sample student drawing:

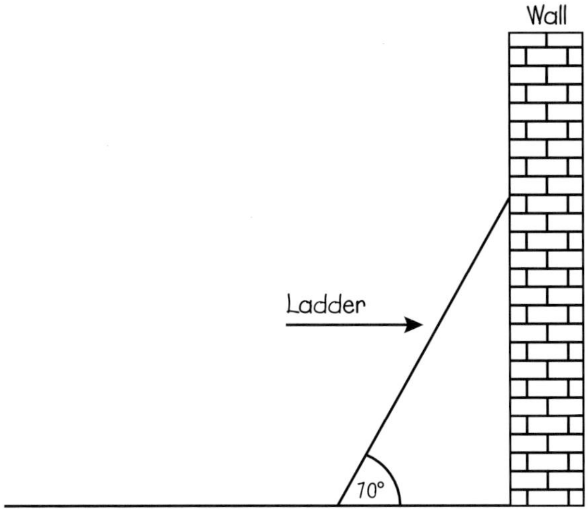

b. Scaled drawings will vary. Sample drawing:

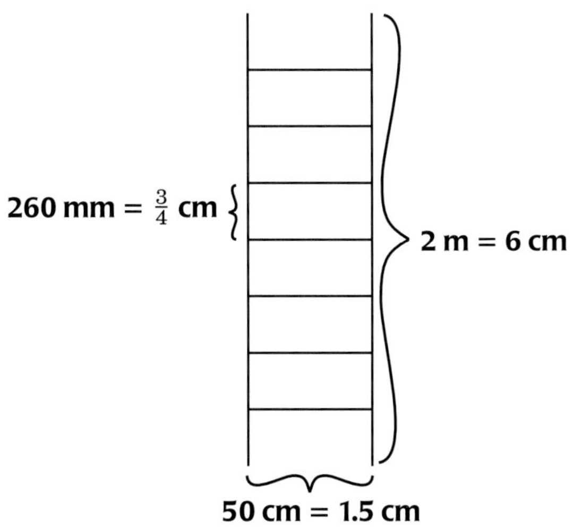

Scale: 1 meter = 3 centimeters

c. Answers will vary. Sample responses:

The classroom doorway is about as tall as the ladder.

The height of a bookcase is about the same height as the ladder.

Hints and Comments

Materials Made to Measure Problems assessment, page 96 of the Teacher Guide (one per student); centimeter rulers (one per student); calculators (one per student); compass cards from Transparency Master 1 or protractors (one per student); grid paper, optional (one sheet per student)

Overview Students examine a drawing of an ankle and determine the angle formed by the ankle's position. They also draw a ladder that forms a 70° angle with the horizontal plane and make a scale drawing of another ladder using given dimensions.

About the Mathematics Problem **4** assesses students' ability to find (compute, measure and draw) angles using standard units and to understand and use relationships between length and angle.

Problem **5** assesses their ability to estimate and find (compute, measure, and draw) lengths using standard units; to use simple relationships between and within measurement systems; to develop and use a frame of reference for measurement to solve problems; and to select and use appropriate measurement units.

Planning Be sure that students have compass cards or protractors available to use in measuring and drawing the given angles. You may want students to work individually on these assessment problems.

Comments about the Problems

4. b. Encourage students to position their own ankles at the suggested angle in order to help them answer this problem.

5. b. Remind students to indicate the scale used in their drawings of the ladder.

MADE TO MEASURE ASSESSMENT PROJECT

Use additional paper as needed.

Use historical, metric, and/or English measurement units to measure the lengths, widths, surface areas, volumes, and, if possible, angles of your head, neck, chest, arms, legs, hands, and total body. Write a paragraph that describes how you calculated each measurement. You might want to make a poster by drawing a picture of yourself that shows your findings.

Solutions and Samples
of student work

Answers for the lengths, widths, surface areas, volumes, and angles of various body parts will vary. Sample student response for hand measurements:

I measured my hand using a centimeter ruler. The length of my hand is 18.5 centimeters (from the tip of my middle finger to the end of my palm), and the width of my hand is 9 centimeters at it's widest point.

To estimate the area of my hand, I traced the outline of my hand on a piece of centimeter grid paper. Then I counted all of the squares and half squares within the drawing. I found the area of my hand to be about 126 cm^2.

I estimated the surface area of my hand by multiplying the area of my hand (126 cm^2) by 2 and adding 30 cm^2 to represent the sides:

$126\ cm^2 \times 2 = 252\ cm^2 + 30\ cm^2 = 282\ cm^2$

I estimated the volume of my hand by multiplying the surface area of one side (126 cm^2) by the thickness of my hand (about 1.5 cm):

$126\ cm^2 \times 1.5\ cm = 189\ cm^3$

Hints and Comments

Materials Made to Measure Project assessment, page 97 of the Teacher Guide (one per student); centimeter and inch rulers (one per student); compass cards or protractors (one per student); grid paper, optional (one sheet per student); calculators (one per student)

Overview Students use nonstandard and standard measurement units to measure the length, width, surface area, volume, and angle of their body parts. They present their findings in a written report and/or poster.

About the Mathematics This project assesses all of the major goals of the *Made to Measure* unit. This assessment tool may be used in place of the five assessment problems provided. Since this project differs somewhat in format from the more structured assessment problems, you may want to consider using the following criteria in scoring students' finished projects:

- The written or oral report should include a complete and clear description of how students measured or calculated the lengths, areas, volumes, and angles of their body parts
- Students should use appropriate measurement units.
- Students' measurement results should be reasonable.
- Consider the number of measurements made by students in determining their score.
- The written or oral report should include some mention of the relationships among the various standard and nonstandard measurement units used.
- The creativity students used in making their measurements and/or preparing their reports/posters and the neatness and organization of the reports should receive some consideration.

Be sure that students know the criteria on which they will be judged before they begin this assessment project.

Planning You may want to use this assessment as a long-range project that students work on at home. You might assign it about one week before the end of the unit and have students share their written reports and posters with the class upon the unit's completion.

Made to Measure

Glossary

acre (p. 44) a unit of area used in the U.S. and England equal to 160 square rods, 4,840 square yards, or 4,047 square meters

cord (p. 62) a unit of wood cut for fuel equal to a stack 4 feet × 4 feet × 8 feet, or 128 cubic feet

cubit (p. 62) an ancient unit of measure for length equal to the distance from a person's elbow to the tip of the extended middle finger

double pace (p. 20) the distance from the back of one foot to the front of the same foot after two steps have been taken

ergonomics (p. 68) the study of people and their work environments. One goal of ergonomics is to design equipment and machines for safe and efficient use

face cord (p. 62) a stack of firewood 4 feet × 4 feet × 16 inches, or $42\frac{2}{3}$ cubic feet

fathom (p. 6) the distance from the middle fingertip of one hand to the middle fingertip of the other hand of a person holding his or her arms fully extended sideways. The fathom is a unit of length, now standardized to 6 feet, used especially for measuring the depth of water

foot (p. 6) the distance from the heel to the toe. This measurement is now standardized to 12 inches

full cord (p. 62) a stack of firewood 4 feet × 4 feet × 8 feet, or 128 cubic feet

furlong (p. 20) a unit of distance equal to 220 yards

hand span (p. 6) the distance from the tip of the thumb to the tip of the little finger when the fingers are spread out as far as possible

metric system (p. 14) a system of measurement in which the basic units are the following: length—the meter, mass—the gram, and volume—the liter. Other units in the metric system are related to the basic units by powers of 10

mile (p. 20) the Roman mile was 1,478.7 meters long; This measurement is now standardized to be 5,280 feet.

nomogram (p. 42) a graphic representation that consists of several lines marked off to scale and arranged in such a way that, by using a straightedge to connect known values on two lines (the height and the weight of a person), an unknown value (body surface area) can be read at the point of intersection with a third line

pace (p. 6) the distance from the back heel of one foot to the front toe of the other foot for a normal step

rood (p. 44) a British unit of land area equal to $\frac{1}{4}$ acre; also, a British unit of length equal to 7 or 8 yards or sometimes a rod (5.5 yards)

royal cubit (p. 62) an ancient Egyptian standard length which measures 52.4 centimeters, or 20.62 inches

Scottish thumb (p. 12) the mean distance across the thumb of the thumbs of three men: a large man, an ordinary-size man, and a small man. The Scottish thumb, or standard thumb, is now called an inch

statute mile (p. 20) from British law, 5,280 feet

stitch (p. 16) one-third of a standard thumb

surface area (p. 32) the area of all of the surface or material that covers a body

yard (p. 6) the distance from the nose to the fingertips when the arm is held out to the side. This distance is now standardized to 3 feet.

Blackline Masters

Dear Family,

Your child will soon begin the *Mathematics in Context* unit *Made to Measure.* This unit is all about the measurement of lengths, areas, volumes, and angles. These topics are a part of the study of geometry.

In the unit, students measure the lengths of their feet, hands, and thumbs and look for relationships between them. They measure the areas of boxes, a football field, and even their own bodies. They measure volumes of a deck of cards, bottles, and stacks of firewood. They also measure the angles made by arms and wrists.

To help your child, get him or her involved in measuring around the house. Ask your child to measure ingredients in recipes, fabric for sewing, the length of a hall for carpet, or the dimensions of the sheets needed for a bed. Compare your child's height with the heights of other members of the family.

Discuss situations in which you might estimate a measurement. Have your child estimate, for example, how many pennies are in a jar, how long it will take for something to cook, or how long a log will burn in the fireplace.

By showing children how measurement is used daily, they will better appreciate what they are studying in school.

Sincerely,

The Mathematics in Context Development Team

Dear Student,

You will soon be starting the *Mathematics in Context* unit *Made to Measure.* The unit is all about measuring: measuring your feet, hands, and thumbs; measuring wood, a football field, and a deck of cards; and measuring the angles made by your arm and wrist. You might be amazed by what you can measure!

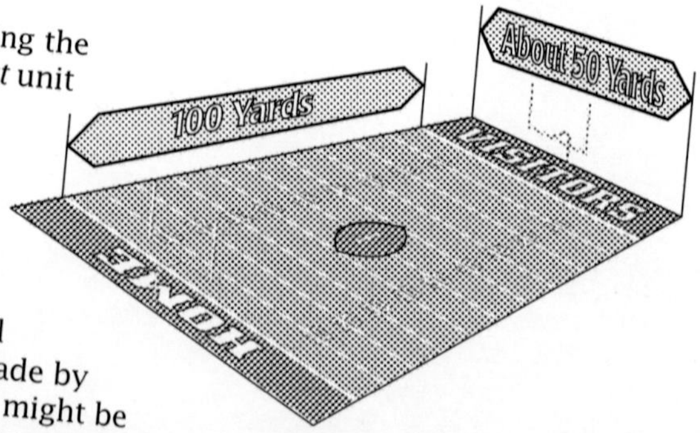

You will find that mathematics plays an important role in measurement. Every time you measure something, you might ask yourself the following questions:

- Will every person measuring this item get the same measurement that I did?
- Do all of these things have the same measurement?
- What other units of measure can I use?
- Are there other ways to measure these things?

Whenever you make a measurement in this unit, picture how big—or small or steep or short—that measurement is. When you can do this with all of the measurements in this unit, you are well on your way to becoming a mathematician!

Sincerely,

The Mathematics in Context Development Team

Name________________________________

Transparency Master 1

Use with *Made to Measure*, pages 31–34 and 37.

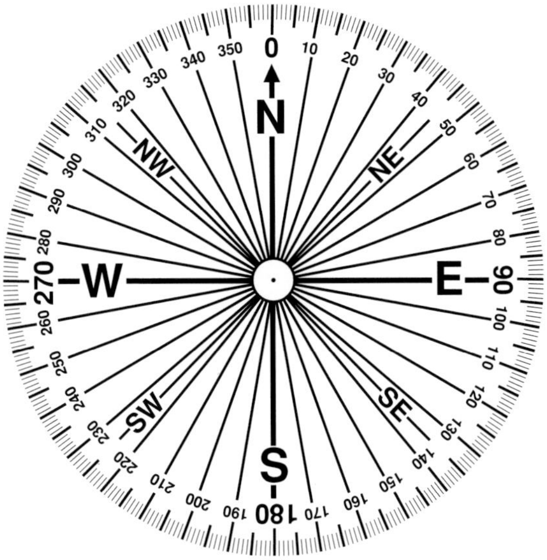

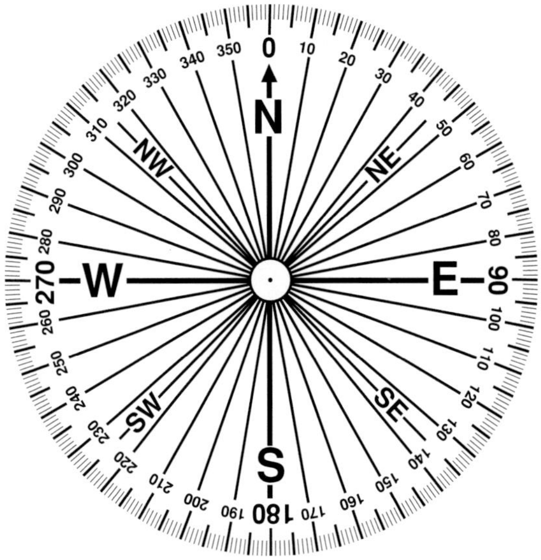

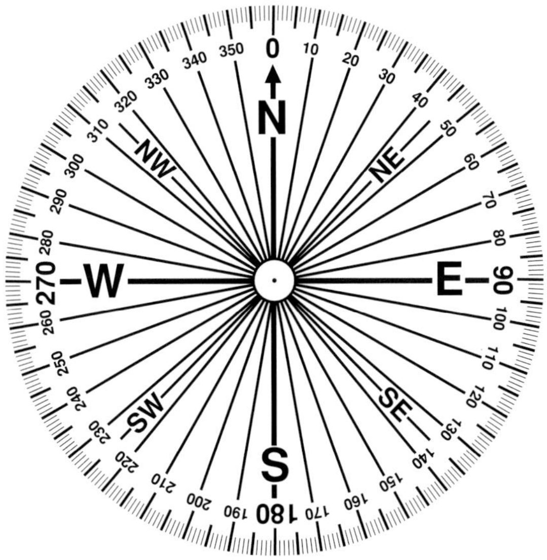

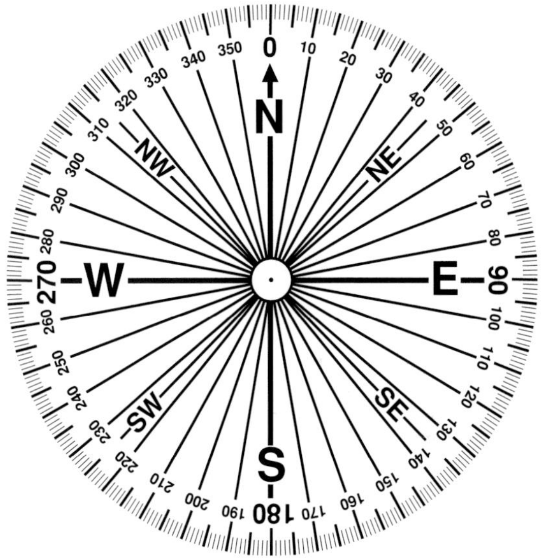

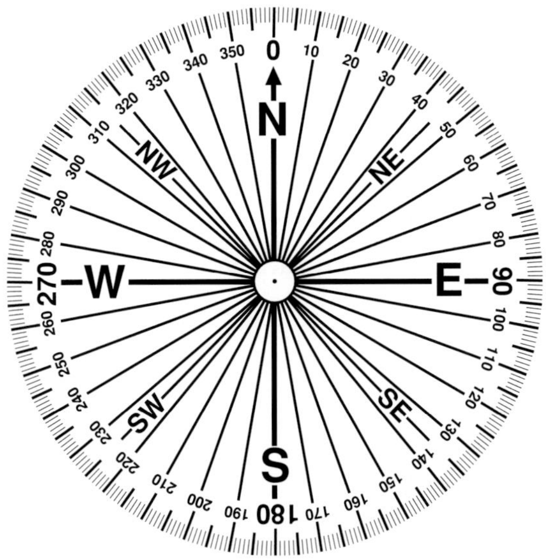

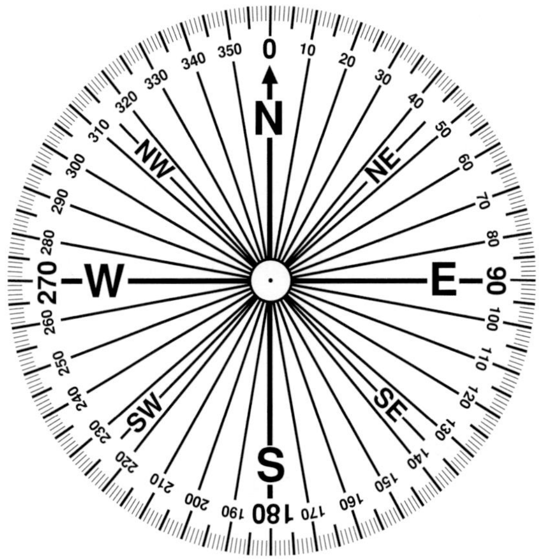

Name ______________________ Date ______________________

MADE TO MEASURE PROBLEMS

Use additional paper as needed.

1. **a.** Use one or more historical measurement units that are related to the human body to find the length and width of this sheet of paper.

   b. Explain why you chose the measurement unit(s) you did.

   c. Explain why it is not likely that everyone in your class got the same answer for part **a.**

When measured in centimeters, this paper is about 21 centimeters wide and 28 centimeters long.

   d. Write two statements that show the relationship between the historical measurement unit(s) you used in part **a** and centimeters.

   e. Estimate the area of this paper in historical measurement units *and* in square centimeters.

Suppose you make a cube out of this sheet of paper by using a net similar to the one shown below. Suppose the net is large enough to just fit this sheet of paper.

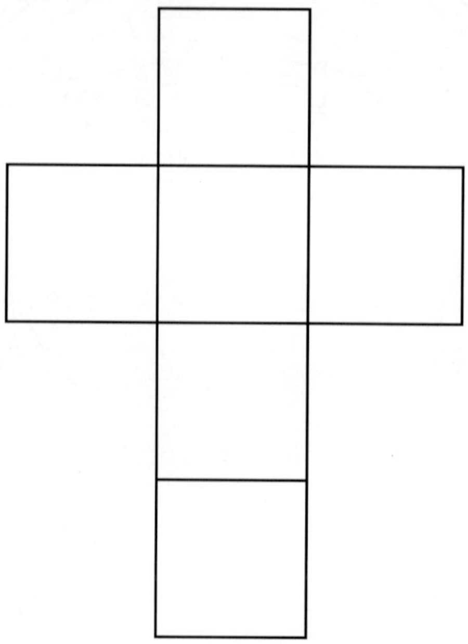

   f. Will this cube hold more or less than one liter? Explain your answer.

Name ________________________ Date ________________________

Use additional paper as needed.

In a survey, people were asked how far they commute daily from home to their place of work. Here are some of their responses:

Jody said, "I commute about three miles to get to work every day."

Carl said, "It takes me about 15 minutes to get to work."

Brenda said, "I live a little less than 5 kilometers from work."

Edward said, "It takes me about 10 minutes to get to work by car."

2. a. Of the four people mentioned above, who do you think lives the farthest from his or her place of work? Explain your reasoning.

b. Which two of these people live the same distance from their place of work? Explain.

Hannah has a dog named Fido. Shown below is a picture of Fido and some of Fido's body measurements.

- Body height: 36 centimeters
- Height of legs: 30 centimeters
- Body length: 44 centimeters
- Circumference around body: 45 centimeters
- Head length: 18 centimeters

Hannah has taken some measurements of Fido, which are shown above. Hannah wants to know what Fido's surface (skin) area is.

3. a. Describe at least two ways that Hannah could estimate Fido's surface area.

b. Estimate Fido's surface area. Explain how you made your estimation.

c. Estimate Fido's volume in cubic decimeters.

Name ____________________ Date ____________________

Use additional paper as needed.

The drawing below shows a person's ankle stretching as far up as it can go.

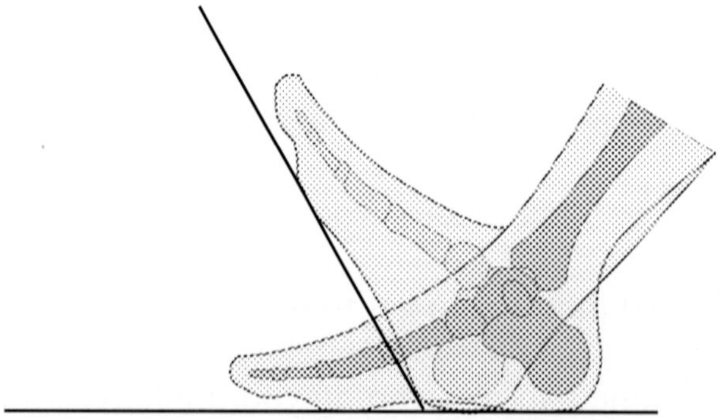

4. a. Measure the outer angle formed by the two positions of the ankle.

b. Which of the two positions looks the most comfortable for an ankle. Explain why.

"Climbing ladders is most convenient when the ladder is at an angle of 70° and the rungs are 260 millimeters apart." E. Grandjean

5. a. The picture on the right shows the side view of a wall. Draw a ladder against it that is at an angle of 70 degrees with the ground.

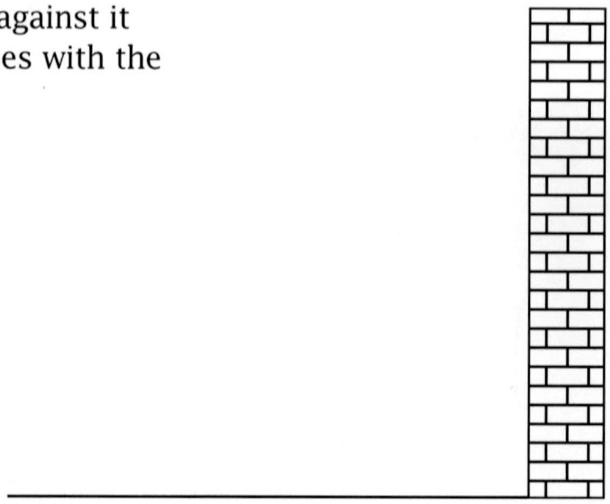

b. On the back of this paper, draw a ladder that is 2 meters long and 50 centimeters wide to scale. Remember to draw the rungs 260 millimeters apart.

c. Find an object in your classroom that is about as tall as the ladder you drew.

Name ______________________________ Date ______________________________

MADE TO MEASURE ASSESSMENT PROJECT

Use additional paper as needed.

Use historical, metric, and/or English measurement units to measure the lengths, widths, surface areas, volumes, and, if possible, angles of your head, neck, chest, arms, legs, hands, and total body. Write a paragraph that describes how you calculated each measurement. You might want to make a poster by drawing a picture of yourself that shows your findings.

Section A. Lengths

1. Answers will vary. Sample responses:

a. the size of a person's head, especially from the ear to the chin; the size of a person's hand

b. the height of a child; the length of a child's legs; the width of a child's body

c. the height of an average adult person; the length of his or her arms

2. Answers will vary. Sample responses:

thumb, hand span, foot, yard, pace, fathom, and cubit

Sample definitions:

Thumb: the width of an average man's thumb;

Hand span: the width of an outstretched hand from the tip of the last finger to the tip of the thumb;

Foot: the length of an average man's foot;

Yard: the length of a man's reach from his face to the tip of his finger;

Pace: the distance from the back heel of one foot to the front toe of the opposite foot after one step has been taken;

Fathom: the distance from the middle fingertip of one hand to the middle fingertip of the other hand of a person holding his or her arms fully extended sideways;

Cubit: the distance from a person's elbow to the tip of the extended middle finger.

3. a. feet

b. pace

c. thumb

4. a. U.K. shoe size 5; 30 stitches

b. U.K. shoe size 3; 29 stitches

c. U.K. shoe size 1; 26 stitches

5. Answers and explanations will vary. Sample responses:

a. I would express the distance from home to school in time units, because I know it takes me 10 minutes to walk to school but I'm not sure what the actual distance is.

b. I would want to know the distance from my city to New York City in miles so I could get an idea about how far away it is from my city.

c. I would express the distance from start to finish of a hiking trail using time units, especially if it is a hiking trail in the mountains. The actual trail distance may be short, but if the terrain is rocky or steep, it might take a long time to hike it.

Section B. Areas

1. Estimates will vary. Sample estimates:

a. 1,800 cm^2. Strategies will vary. Sample strategy:

I wrapped a basketball in newspaper and made sure that no newspaper overlapped. Then I removed the newspaper and rearranged the pieces into the shape of a rectangle. I measured the length and width of the rectangle with a centimeter ruler and multiplied to find the area.

b. 570 cm^2. Strategies will vary. Sample strategy:

I measured the length, width, and depth of the book:

length: 18 cm; width: 11 cm; depth: 3 cm.

Then I estimated the area of the front of the book and doubled it:

11 cm × 18 cm is about 10 × 20 = 200 cm^2 × 2 = 400 cm^2

Then I estimated the area of the book's spine and doubled it:

3 cm × 18 cm is about 3 × 20 = 60 cm^2 × 2 = 120 cm^2

Then I estimated the area of the top of the book and doubled it:

3 cm × 11 cm is about 3 × 10 = 30 cm^2 × 2 = 60 cm^2

So, I estimated the total surface area of the book to be:

400 cm^2 + 120 cm^2 + 60 cm^2 = 580 cm^2

c. 1,656 cm^2 (for a cereal box with dimensions 6 cm × 18 cm × 30 cm). Strategies will vary. Students may use the same strategy as illustrated in part **b.**

2. Answers will vary. Sample response:

It was easy to estimate the surface area of the cereal box and the book, since I could measure the dimensions of each object using a ruler. I could not measure the dimensions of the ball and estimate its surface area using the same method, since a ball has a curved surface.

3. a. Formulas will vary. Sample formula:

Surface area = circumference of body (at the hips) × *height* + $2\pi r^2$

Explanations will vary. Sample explanation:

I thought of the torso and the legs of my body as the rectangular flat surface of a cylinder's wall. The height of the cylinder is the same as my height. The length of the cylinder's wall is the same as the distance around my body at my hips (the circumference). So, you multiply those two dimensions to find the area: circumference of body (at the hips) × height.

Then I needed to find a way to estimate the areas of the two "circles" that form the top and bottom of the cylinder (my head and feet). To find the diameter, I divided the circumference of my hips by π and then divided the diameter in half to find the radius. Then I found the area of the two circles by multiplying $2\pi r^2$.

b. Answers will vary. Sample response: about 17,500 cm^2

Using the formula in part **a:** Circumference of the hips is 90 cm; the height of the person is 180 cm; and the radius of the hips is about 90 cm: 2 × 3.14 × 14.3 × 14.3 + 90 × 180 = 1,284 + 16,200 = 17,484 cm^2.

4. a. about 14 football fields. Strategies will vary. Sample strategy:

One acre is the same as 43,560 sq ft, so I used a ratio table to estimate that 15 acres ≈ 654,000 sq ft.

×3 ×5

Acres	1	3	15
Square Feet	43,600	130,800	654,000

One football field is 42,000 sq ft (300 ft × 140 ft = 42,000 sq ft). So, I used a ratio table to estimate how many football fields would equal 654,000 sq ft:

×2 ×5 ÷2

Football Fields	1	2	10	5	15
Square Feet	42,000	84,000	420,000	210,000	630,000

420,000 + 210,000

b. Answers will vary. Sample response:

600 feet by 1,090 ft. I used a guess-and-check strategy with my calculator until I found two numbers that I multiplied together to make a product of 654,000.

Section C. Volumes

1. Answers will vary. Sample responses:

One liter equals 1 dm^3, or 1,000 cm^3.
One gallon is about 4 liters, or 4 quarts or 4 dm^3.
One quart equals about 1 liter.
Two pints equal 1 quart.

2. Answers will vary. Sample responses:

a. 3,240 cm^3. This is the volume of a cereal box that measures 6 cm by 18 cm by 30 cm ($6 \times 18 \times 30 = 3{,}240\ cm^3$).

b. 18 cm^3. This is the volume of a package of gum that measures 1 cm by 3 cm by 6 cm ($1 \times 3 \times 6 = 18\ cm^3$).

3. Answers will vary. Sample response:

The flower box could be 10 cm by 10 cm by 30 cm. This container has a volume of 3,000 m^3 ($10 \times 10 \times 30 = 3{,}000\ m^3$).

4. a. 960 cubic centimeters. Strategies will vary. Sample strategy:

I used the volume formula: volume = area of the base × height: $80\ cm^2 \times 12\ cm = 960\ cm^3$

b. 360 cubic centimeters. Strategies will vary. Sample strategy:

I used the volume formula: *volume* = *length* × *width* × *height:* $4\ cm \times 15\ cm \times 6\ cm = 360\ cm^3$

5. Estimates will vary. Sample estimates:

a. about 160 liters

b. about 1 liter

c. about $\frac{1}{2}$ liter

6. a. Estimates will vary. Sample estimate: 50 blocks

b. I used the 4 cm × 15 cm × 6 cm blocks from problem **4b.**

c. Explanations will vary. Sample explanation:

Small cubes will be more accurate than large blocks in measuring the volume of the human body, which is made up of different shapes and small parts.

Section D. Angles

1.

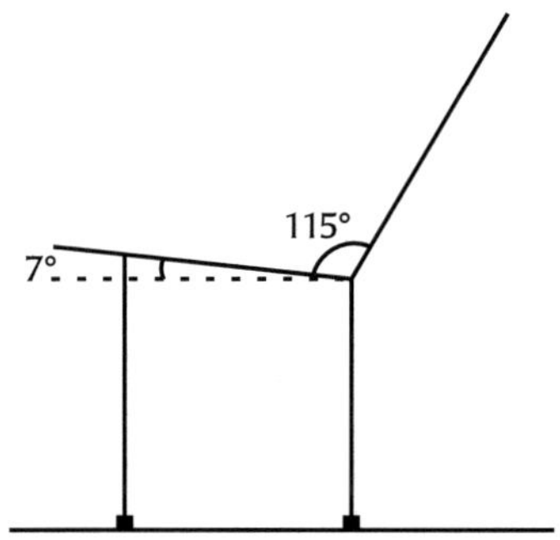

2. a. The angle is 8°.

b. The angle is 7°.

CREDITS

Cover

Design by Ralph Paquet/Encyclopædia Britannica Educational Corporation.

Collage by Koorosh Jamalpur/KJ Graphics.

Title Page

Phil Geib/Encyclopædia Britannica Educational Corporation.

Illustrations

6, 8 (a, b, c, e, f) Phil Geib/Encyclopædia Britannica Educational Corporation; **8 (d), 10** Tom Zielinski/ Encyclopædia Britannica Educational Corporation; **12, 16 (top)** Phil Geib/ Encyclopædia Britannica Educational Corporation; **16 (bottom)** Tom Zielinski/Encyclopædia Britannica Educational Corporation; **18, 20, 22 (top)** Phil Geib/Encyclopædia Britannica Educational Corporation; **22 (bottom)** Phil Geib and Tom Zielinski/Encyclopædia Britannica Educational Corporation; **24, 26, 28, 32** Phil Geib/ Encyclopædia Britannica Educational Corporation; **34** Phil Geib and Tom Zielinski/Encyclopædia Britannica Educational Corporation; **36, 38 (top)** Phil Geib/Encyclopædia Britannica Educational Corporation; **38 (bottom), 40, 42** Phil Geib and Tom Zielinski/Encyclopædia Britannica Educational Corporation; **44, 46, 50, 56 (top)** Phil Geib/ Encyclopædia Britannica Educational Corporation; **56 (bottom)** Phil Geib and Tom Zielinski/ Encyclopædia Britannica Educational Corporation; **58, 60, 62, 64, 68** Phil Geib/Encyclopædia Britannica Educational Corporation; **70, 72** Phil Geib and Tom Zielinski/Encyclopædia Britannica Educational Corporation; **74, 75, 76, 79, 84, 86** Phil Geib/ Encyclopædia Britannica Educational Corporation.